Advances in Industrial Control

Springer
London
Berlin
Heidelberg
New York
Hong Kong
Milan
Paris
Tokyo

Other titles published in this Series:

Performance Assessment of Control Loops: Theory and Applications
Biao Huang & Sirish L. Shah

Advances in PID Control
Tan Kok Kiong, Wang Qing-Guo & Hang Chang Chieh with Tore J. Hägglund

Advanced Control with Recurrent High-order Neural Networks: Theory and Industrial Applications
George A. Rovithakis & Manolis A. Christodoulou

Structure and Synthesis of PID Controllers
Aniruddha Datta, Ming-Tzu Ho and Shankar P. Bhattacharyya

Data-driven Techniques for Fault Detection and Diagnosis in Chemical Processes
Evan L. Russell, Leo H. Chiang and Richard D. Braatz

Bounded Dynamic Stochastic Systems: Modelling and Control
Hong Wang

Non-linear Model-based Process Control
Rashid M. Ansari and Moses O. Tadé

Identification and Control of Sheet and Film Processes
Andrew P. Featherstone, Jeremy G. VanAntwerp and Richard D. Braatz

Precision Motion Control
Tan Kok Kiong, Lee Tong Heng, Dou Huifang and Huang Sunan

Nonlinear Identification and Control: A Neural Network Approach
Guoping Liu

Digital Controller Implementation and Fragility: A Modern Perspective
Robert S.H. Istepanian and James F. Whidborne

Optimisation of Industrial Processes at Supervisory Level
Doris Sáez, Aldo Cipriano and Andrzej W. Ordys

Applied Predictive Control
Huang Sunan, Tan Kok Kiong and Lee Tong Heng

Hard Disk Drive Servo Systems
Ben M. Chen, Tong H. Lee and Venkatakrishnan Venkataramanan

Robust Control of Diesel Ship Propulsion
Nikolaos Xiros

Hydraulic Servo-systems
Moheiddine Jelali and Andreas Kroll

Model-based Fault Diagnosis in Dynamic Systems Using Identification Techniques
Silvio Simani, Cesare Fantuzzi and Ron J. Patton

Freddy Garces, Victor M. Becerra,
Chandrasekhar Kambhampati and Kevin Warwick

Strategies for Feedback Linearisation

A Dynamic Neural Network Approach

With 57 Figures

 Springer

Freddy Garces, MSc, PhD
Victor M. Becerra, MSc, PhD
Department of Cybernetics, University of Reading, UK

Chandrasekhar Kambhampati, PhD
Department of Computer Science, The University of Hull, UK

Kevin Warwick, DSc, DrSc, PhD
Department of Cybernetics, University of Reading, UK

British Library Cataloguing in Publication Data
Strategies for feedback linearisation : a dynamic neural
   network approach. - (Advances in industrial control)
   1.Feedback control systems - Mathematical models
   2.Nonlinear control theory 3.Neural networks (Computer
   Science)
   I.Garces, Freddy
   629.8'312
   ISBN 1852335017

Library of Congress Cataloging-in-Publication Data
A catalog record for this book is available from the Library of Congress.

Apart from any fair dealing for the purposes of research or private study, or criticism or review, as permitted under the Copyright, Designs and Patents Act 1988, this publication may only be reproduced, stored or transmitted, in any form or by any means, with the prior permission in writing of the publishers, or in the case of reprographic reproduction in accordance with the terms of licences issued by the Copyright Licensing Agency. Enquiries concerning reproduction outside those terms should be sent to the publishers.

ISBN 1-85233-501-7 Springer-Verlag London Berlin Heidelberg
a member of BertelsmannSpringer Science+Business Media GmbH
http://www.springer.co.uk

© Springer-Verlag London Limited 2003
Printed in Great Britain

MATLAB® and SIMULINK® are the registered trademarks of The MathWorks Inc., 3 Apple Hill Drive Natick, MA 01760-2098, U.S.A. http://www.mathworks.com

Mathematica® is the registered trademark of Wolfram Research, Inc., 100 Trade Centre Drive, Champaign, IL 61820-7237, USA, http://www.wolfram.com.

Maple® is the registered trademark of Waterloo Maple, Inc., 57 Erb Street, W. Waterloo, Ontario, Canada, N2L 6C2. http://www.maplesoft.com.

The use of registered names, trademarks etc. in this publication does not imply, even in the absence of a specific statement, that such names are exempt from the relevant laws and regulations and therefore free for general use.

The publisher makes no representation, express or implied, with regard to the accuracy of the information contained in this book and cannot accept any legal responsibility or liability for any errors or omissions that may be made.

Typesetting: Electronic text files prepared by authors
Printed and bound by Athenæum Press Ltd., Gateshead, Tyne & Wear
69/3830-543210 Printed on acid-free paper SPIN 10836576

Advances in Industrial Control

Series Editors

Professor Michael J. Grimble, Professor of Industrial Systems and Director
Professor Michael A. Johnson, Professor of Control Systems and Deputy Director

Industrial Control Centre
Department of Electronic and Electrical Engineering
University of Strathclyde
Graham Hills Building
50 George Street
Glasgow G1 1QE
United Kingdom

Series Advisory Board

Professor E.F. Camacho
Escuela Superior de Ingenieros
Universidad de Sevilla
Camino de los Descobrimientos s/n
41092 Sevilla
Spain

Professor S. Engell
Lehrstuhl für Anlagensteuerungstechnik
Fachbereich Chemietechnik
Universität Dortmund
44221 Dortmund
Germany

Professor G. Goodwin
Department of Electrical and Computer Engineering
The University of Newcastle
Callaghan
NSW 2308
Australia

Professor T.J. Harris
Department of Chemical Engineering
Queen's University
Kingston, Ontario
K7L 3N6
Canada

Professor T.H. Lee
Department of Electrical Engineering
National University of Singapore
4 Engineering Drive 3
Singapore 117576

Professor Emeritus O.P. Malik
Department of Electrical and Computer Engineering
University of Calgary
2500, University Drive, NW
Calgary
Alberta
T2N 1N4
Canada

Professor K.-F. Man
Electronic Engineering Department
City University of Hong Kong
Tat Chee Avenue
Kowloon
Hong Kong

Professor G. Olsson
Department of Industrial Electrical Engineering and Automation
Lund Institute of Technology
Box 118
S-221 00 Lund
Sweden

Professor A. Ray
Pennsylvania State University
Department of Mechanical Engineering
0329 Reber Building
University Park
PA 16802
USA

Professor D.E. Seborg
Chemical Engineering
3335 Engineering II
University of California Santa Barbara
Santa Barbara
CA 93106
USA

Doctor I. Yamamoto
Technical Headquarters
Nagasaki Research & Development Center
Mitsubishi Heavy Industries Ltd
5-717-1, Fukahori-Machi
Nagasaki 851-0392
Japan

SERIES EDITORS' FOREWORD

The series *Advances in Industrial Control* aims to report and encourage technology transfer in control engineering. The rapid development of control technology has an impact on all areas of the control discipline. New theory, new controllers, actuators, sensors, new industrial processes, computer methods, new applications, new philosophies…, new challenges. Much of this development work resides in industrial reports, feasibility study papers and the reports of advanced collaborative projects. The series offers an opportunity for researchers to present an extended exposition of such new work in all aspects of industrial control for wider and rapid dissemination.

Nonlinear control methods continue to exert a continuing fascination for current researchers in control systems techniques. Many industrial systems are nonlinear as was so ably demonstrated in the recent *Advances in Industrial Control* monograph on hydraulic servo-systems by M. Jelali and A. Kroll. However, the need to use a nonlinear control technique depends on the severity of the nonlinearity and the performance specification of the application. In some cases it is imperative that a nonlinear technique be used. The type of technique which is applied usually depends on the available information on the system description. This is the key determinant in the development of new nonlinear control methods. Over the next few years it is hoped that the nonlinear control paradigm will produce several methods which will be easily and widely applicable in industrial problems. In the meantime the search and development research go on.

This succinct monograph by Freddy Garces, Victor Becerra, Chandrasekhar Kambhampati and Kevin Warwick is a very useful contribution to current explorations of nonlinear control. It uses the feedback linearisation technique but pragmatically it merges it with neural network system models. This is increasingly the flavour of some of the new nonlinear control methods appearing today where a key objective is to harness the techniques to solving typical industrial problems. The monograph has the advantage of presenting just the right amount of material to understand and demonstrate the technique. Three case studies give useful support to the demonstration section of the monograph.

The monograph should be of interest to a wide range of academic and industrial control engineers. The focussed nature of the presentation makes the monograph highly appropriate for self-study or even as a study text for an industrial course. We welcome its addition to the *Advances in Industrial Control* monograph series.

M.J. Grimble and M.A. Johnson
Industrial Control Centre
Glasgow, Scotland, U.K.

PREFACE

Aim of the book. The main goal of this book is to present recent results in the area of feedback linearisation using empirical models based on dynamic neural networks and to provide the reader with methods for analysing, designing and implementing these techniques.

Background. Many common control problems involve controlled systems that exhibit nonlinear behaviour in that the relationships between controlled and manipulated variables depend on the operating conditions. If the nonlinearities are mild or the operating conditions do not change much, then the effect of nonlinearities may not be severe, and linear control techniques are applicable.

However, many industrial systems exhibit highly nonlinear behaviour and they may be required to operate over a wide range of operating conditions. When conventional linear controllers are used to control highly nonlinear processes, the controllers must be tuned in a conservative manner in order to avoid unstable behaviour. However, this can result in a serious deterioration of control performance. Thus, more sophisticated control techniques are required that use information about the nonlinearities of the controlled system.

The last few decades have witnessed a tremendous development in nonlinear control theory. One important nonlinear control technique is known as feedback linearisation. This technique was developed in the 1970s and consists of transforming a nonlinear system into a controllable linear system by means of static state feedback and nonlinear transformations. Undoubtedly, one of the main reasons to study the feedback linearisation problem lies in its potential use in applications. Once a system is feedback linearised, it admits a standard linear controller design. Moreover, systems with multiple inputs and multiple outputs can be linearised and decoupled, allowing an efficient use of single loops with linear controllers.

The information about the nonlinear dynamic behaviour of a system is encapsulated in a dynamic model that often takes the form of a set of nonlinear differential equations. Many nonlinear control techniques are model-based in that they require the use of this type of dynamic model, including the feedback linearisation approach. The direct way of obtaining a nonlinear model of the plant is to derive a physically based model from principles such as mass and energy balances. These models provide a rich physical insight into the process and are applicable over a wide range of operating conditions. However, physical models are often not available due to the engineering effort and cost associated with their development and maintenance. An alternative way to obtain the required model is to identify it using measured input–output data. The development of techniques of system identification has made possible the synthesis of empirical dynamic models using data measured from the system. In recent years, there has been considerable interest in developing nonlinear dynamic models from input–output data and a variety of model structures and techniques are available.

The predominant family of structures for obtaining nonlinear empirical models is known as artificial neural networks, or simply neural networks, which are inspired by the connections of biological neurons. These models consist of a set of interconnected processing units or artificial neurons and are regarded as structures capable of approximating generic nonlinear input–output maps. Inherent capabilities of neural models such as generalisation, parallel distributed processing and nonlinear dynamic approximation make them a promising tool for nonlinear system identification.

The field of neural computations has evolved from neurological roots when the first artificial neural models were proposed to today's solid mathematical formulations. Neural networks can be divided into static and dynamic networks. A static neural model is described by an algebraic equation while a dynamic neural network is represented by a difference or differential equation depending on whether it is based on a discrete or continuous domain, respectively. One of the most common architectures for static neural networks is the multilayer feedforward network or multilayer perceptron. For the purpose of the nonlinear system identification for control, perhaps the most relevant architectures are the ones offering a state space model such as the Hopfield networks and their variations.

A dynamic recurrent neural network, or simply a dynamic neural network, is a collection of dynamic neurons partially interconnected to a function of their own output. Such networks can be represented by a nonlinear state space model. Dynamic recurrent neural networks can approximate a wide range of nonlinear dynamic behaviours and can be considered as generic nonlinear dynamic systems approximators.

Intended readership. This book has been written to serve a wide range of individuals. In order to achieve this, we have attempted to present a balanced view of the theoretical and practical issues. Thus we hope that the book will be of interest both to practising control engineers with an interest in nonlinear control techniques and also to academic researchers in control theory. Case studies are presented to illustrate design and application issues; relevant mathematical proofs are also included. We have made an effort to present intuitive explanations and illustrative examples of the main results discussed in the book. The very nature of nonlinear systems requires the use of some advanced mathematical tools, which are introduced when necessary. We have assumed that the readers have a working knowledge of engineering mathematics and that they have had some exposure to basic linear control theory, including linear state space methods.

Outline of the book. Chapter 1 provides a general outline of the control techniques introduced in the book within the context of nonlinear identification and control theory. Chapter 2 introduces fundamental analytical concepts that will be used later in the book. Chapter 3 presents an introduction to feedback linearisation, including to feedback linearising-decoupling techniques for dynamic systems with multiple inputs and multiple outputs. Chapter 4 presents a general description dynamic neural networks, which are further analysed in terms of their stability. Training methods relevant to system identification and structure selection techniques are also discussed. Chapter 5 presents theories of approximation relevant to static and dynamic neural networks. Chapter 6 provides a description of the design and implementation of the feedback linearising strategy employing dynamic neural networks. The final control scheme is built up within a multiloop proportional+integral (PI) structure. Chapter 7 provides case studies based on laboratory experiments and simulations and compares the performance of the techniques presented in this book with more conventional control strategies.

Resources. Additional supporting material for this book can be found at the following URL: http://www.rdg.ac.uk/~shs99vmb/strategies.

Acknowledgements. The authors would like to thank several people who helped in different ways in the preparation of this book: João Calado, Pedro Silva, Slawomir Nasuto, Kandasamy Pirabakaran and Oliver Jackson. The authors are also grateful with the series Editors for their support. Research funding from the following institutions is gratefully acknowledged: the Engineering and Physical Sciences Research Council (EPSRC), Fundayacucho, the British Council, and the University of Reading.

Reading, UK, August 2002

Freddy Garces

Victor M. Becerra

Kevin Warwick

Hull, UK, August 2002

Chandra Kambhampati

CONTENTS

1. **Introduction** 1
   1.1 The Need for Nonlinear Control in Industrial Processes 1
   1.2 Nonlinear Control Strategies 1
       1.2.1 Gain scheduling 2
       1.2.2 Feedback linearisation 2
       1.2.3 Feedback linearisation-decoupling 4
   1.3 Nonlinear System Models: a Key Issue 4
   1.4 Neural Networks 5
   1.5 System Identification 8
   1.6 Static Neural Networks for Identification and Control 9
   1.7 Dynamic Neural Networks for Nonlinear Identification 10
   1.8 Input–Output Linearisation-Decoupling and the Use of Dynamic Neural Networks 12
   1.9 Potential applications 12

2. **Fundamental Concepts** 15
   2.1 Elementary Concepts of Geometric Theory 15
       2.1.1 Linear vector spaces 15
       2.1.2 Euclidean space 15
       2.1.3 Vector norms 16
       2.1.4 Matrix norms 16
       2.1.5 Sets 17
       2.1.6 Vector fields 17
       2.1.7 Differential operations 18
       2.1.8 Distributions 20
   2.2 Stability of Nonlinear Systems 21
   2.3 Summary 26

3. **Introduction to Feedback Linearisation** 27
   3.1 Nonlinear Control Affine Systems 27
       3.1.1 Relative degree and characteristic matrix 30
       3.1.2 Zero dynamics and normal form 33
       3.1.3 Input–output linearisation 39
       3.1.4 Input–output linearisation and decoupling 46

xiv   Contents

        3.1.5   Stability of input–output linearised systems 48
   3.2   Review of Other Linearisation Techniques 54
        3.2.1   Exact linearisation 54
        3.2.2   Linearisation by immersion 55
        3.2.3   Volterra linearisation 56
   3.3   General Nonlinear Systems 57
        3.3.1   Lie derivative and relative degree 57
        3.3.2   Approximate input–output linearisation 58
   3.4   Symbolic Algebra Software 58
   3.5   Remarks 58
   3.6   Summary 59

4. Dynamic Neural Networks 61
   4.1   Introduction 61
   4.2   Origins of Neural Computation 62
   4.3   Single Layer Neural Network Structure 62
   4.4   Static Multilayer Feedforward Networks 65
   4.5   Dynamic Neural Networks (DNNs) 66
        4.5.1   Vector relative degree of a multi-input multi-output dynamic neural network 68
        4.5.2   Stability of dynamic neural networks 69
   4.6   Training Dynamic Neural Networks 78
        4.6.1   Experiment design and input characterisation 78
        4.6.2   Training as an optimisation problem 79
        4.6.3   Initialising dynamic neural networks 82
        4.6.4   Gradient based optimisation methods 82
        4.6.5   Random search methods 87
   4.7   Validating the Dynamic Neural Models 91
        4.7.1   Overtraining and overfitting 91
        4.7.2   Generalisation, bias and variance 92
        4.7.3   Cross-validation and model structure selection 93
        4.7.4   Regularisation 93
   4.8   A Training Example 94
   4.9   Summary 95

5. Nonlinear System Approximation Using Dynamic Neural Networks 101
   5.1   The Universal Approximation Property of Static Multilayer Networks 101
   5.2   Dynamic Neural Network Structure 102
   5.3   Approximation Ability of Dynamic Neural Networks 103
        5.3.1   Approximation of autonomous nonlinear systems 104
        5.3.2   Approximation of non-autonomous nonlinear systems . 109
        5.3.3   Upper bound on the approximation error of general nonlinear systems 114

| | 5.4 | Summary .... 119 |

6.	**Feedback Linearisation Using Dynamic Neural Networks** . 121
	6.1	Approximate Input–Output Linearisation of Control Affine Systems 122
		6.1.1	Approximate input–output linearisation-decoupling and external control 125
		6.1.2	Stability analysis 128
	6.2	Approximate Input–Output Linearisation for General Nonlinear Systems 130
	6.3	Related Work 131
	6.4	Summary 133

7.	**Case Studies** 135
	7.1	The Pressure Pilot Plant 135
		7.1.1	Description of the plant 135
		7.1.2	Identification results 136
		7.1.3	Globally linearising control experimental results 137
	7.2	Single Link Manipulator 138
	7.3	Evaporator System 141
		7.3.1	Single loop control simulations on the evaporator 151
		7.3.2	Approximate feedback linearisation-decoupling of the evaporator 153
	7.4	Summary 160

References 161

Index 169

CHAPTER 1
INTRODUCTION

1.1 The Need for Nonlinear Control in Industrial Processes

The objective of control is to influence the behaviour of systems. Two important control problems are regulation and tracking. Regulation involves keeping system's variables at desired constant values, while tracking involves forcing them to follow prescribed trajectories. The control problem involves determining the values of the manipulated input using all available information to achieve the control objective.

Most physical systems are nonlinear and multivariable. By nature, they have inherent interconnected nonlinearities in their dynamics where the relationship between the input and output variables varies depending on the operating conditions. For example, for a step change on one of the inputs of the system, parameters such as steady-state gains, time constants and time delays for the outputs depend not only on the amplitude of the step but also on the operating values of the rest of the variables.

Many common control problems involve dynamic systems that exhibit nonlinear behaviour. If the nonlinearities are mild or the operating conditions do not change much, then the effect of nonlinearities may not be severe, and linear control techniques are applicable. However, many industrial systems exhibit strong nonlinear behaviour and they may be required to operate over a wide range of operating conditions. When conventional linear controllers are used to control highly nonlinear systems, the controllers must be tuned in a conservative manner in order to avoid unstable behaviour. However, this can result in a serious deterioration of control performance.

1.2 Nonlinear Control Strategies

An exhaustive review of the nonlinear control theory is very ambitious and is beyond the scope of this book. The area is extremely diverse and undergoing continuous development. This section narrows the focus to the more relevant techniques for the purposes of this book, namely gain scheduling and feedback linearisation.

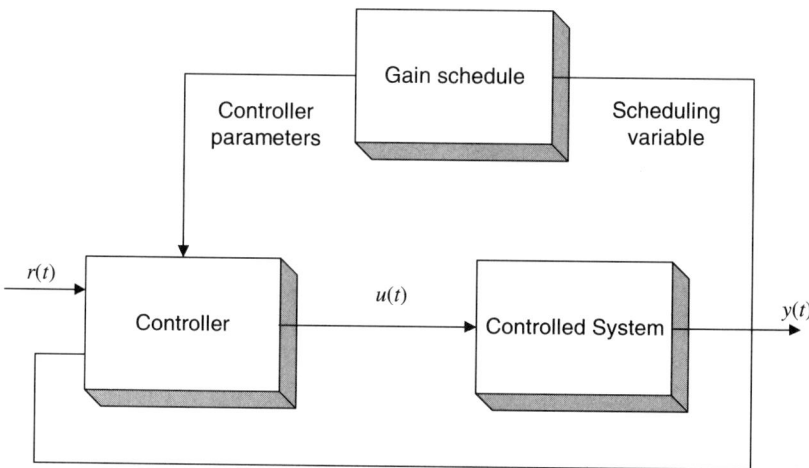

Fig. 1.1. Illustration of gain scheduling

1.2.1 Gain scheduling

Gain scheduling is an engineering approach that has been widely used to compensate for nonlinear process characteristics in single-input single-output (SISO) systems [1, 2, 3]. In gain scheduling, the controller parameters are changed or scheduled following known static nonlinearities of the plant, such that the loop gain is approximately constant. Gain scheduling may be implemented using look-up tables or nonlinear transformations. This technique has been widely applied in different fields, such as flight control and process control. One of the disadvantages of gain scheduling is the lack of a systematic procedure for selecting the scheduling variables. The approach is illustrated in Figure 1.1. The use of nonlinear transformations in gain scheduling is related to the feedback linearization approach [4], which is introduced below.

1.2.2 Feedback linearisation

Jacobian linearisation involves approximating a nonlinear system by a linear one in the vicinity of a reference equilibrium point. In particular, a system described by:

$$\dot{x}(t) = f(x(t), u(t))$$
$$y(t) = h(x(t)) \qquad (1.1)$$

where $f(\cdot, \cdot)$ and $h(\cdot)$ are differentiable nonlinear vector functions, $x \in \Re^n$ is the state, $y \in \Re^p$ is the output and $u \in \Re^m$ is the input, can be locally approximated around an equilibrium point given by (x_s, u_s, y_s) as follows:

$$\frac{d\bar{x}(t)}{dt} = \left[\frac{\partial f}{\partial x}\bigg|_{x_s,u_s}\right]\bar{x} + \left[\frac{\partial f}{\partial u}\bigg|_{x_s,u_s}\right]\bar{u} \qquad (1.2)$$

$$\bar{y}(t) = \left[\frac{\partial h}{\partial x}\bigg|_{x_s}\right]\bar{x}$$

where $\bar{x} = x - x_s$, $\bar{u} = u - u_s$, and $\bar{y} = y - y_s$ are deviations from the equilibrium state, input and output, respectively.

In nonlinear control theory, the term *feedback linearisation* has a very different meaning from Jacobian linearisation. Feedback linearisation is perhaps the most important nonlinear control design strategy developed during the last few decades [5]. It has attracted a great deal of research interest resulting in a rigorously formalised field. The main objective of the approach is to algebraically transform nonlinear system dynamics into linear ones by using state feedback and a nonlinear coordinate transformation based on a differential geometric analysis of the system. By eliminating nonlinearities in the system, conventional linear control techniques can be applied. The linearisation is carried out by model-based state transformations and feedback rather than by linear approximations of the dynamics, as used in Jacobian linearisation where the resulting linear model is only locally valid. Differential geometry has proved to be a successful means of analysing and designing nonlinear control systems, equivalently to that of linear algebra and Laplace transform in relation to linear systems. Feedback linearisation is a strong research field with rigorous mathematical formulations [6-10].

To illustrate the basics of feedback linearisation, consider a second order nonlinear SISO system described by the following state equations:

$$\begin{aligned}\dot{x}_1(t) &= x_2(t) \\ \dot{x}_2(t) &= f_2(x(t)) + g_2(x(t))u(t) \\ y(t) &= x_1(t)\end{aligned} \qquad (1.3)$$

where $f_2(\cdot)$ and $g_2(\cdot)$ are known nonlinear functions and $x(t) = [\,x_1(t)\ x_2(t)\,]^T$. Assume that $g_2(x(t)) \neq 0$ and let:

$$u(t) = \frac{v(t) - f_2(x(t))}{g_2(x(t))} \qquad (1.4)$$

where $v(t)$ is an artificial input variable. Replacing 1.4 in 1.3:

$$\begin{aligned}\dot{x}_1(t) &= x_2(t) \\ \dot{x}_2(t) &= f_2(x(t)) + g_2(x(t)) \times \left[\frac{v(t)-f_2(x(t))}{g_2(x(t))}\right] \\ y(t) &= x_1(t)\end{aligned} \qquad (1.5)$$

In this way, the original nonlinear system 1.3 has been transformed into the following linear system that uses the artificial input variable $v(t)$:

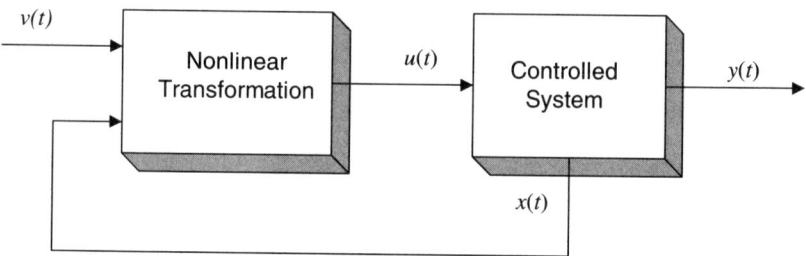

Fig. 1.2. Illustration of feedback linearisation

$$\begin{aligned}\dot{x}_1(t) &= x_2(t)\\ \dot{x}_2(t) &= v(t)\\ y(t) &= x_1(t)\end{aligned} \qquad (1.6)$$

Note that in order to compute the input $u(t)$ as given by Equation 1.4, information about the state $x(t)$ is required, so that state feedback is being employed as is illustrated in Figure 1.2. This is why the approach is known as *feedback linearisation*.

1.2.3 Feedback linearisation-decoupling

For multivariable nonlinear systems, feedback linearisation may be used not only to eliminate nonlinearities in the input–output relations but also to cancel the interactions between variables. A state feedback linearising law is designed to compensate for these interconnections in order to decompose the multivariable nonlinear system into several single-input single-output linear systems [6, 11, 12].

1.3 Nonlinear System Models: a Key Issue

The information about the nonlinear dynamic behaviour of a system is encapsulated in a dynamic model that often takes the form of a set of nonlinear differential equations plus a measurement model:

$$\begin{aligned}\dot{x} &= f(x(t), u(t))\\ y &= h(x(t))\end{aligned} \qquad (1.7)$$

where $f : \Re^n \times \Re^m \to \Re^n$ is a nonlinear mapping, $x \in \Re^n$ is a state vector, $u \in \Re^m$ is a vector of input variables, $y \in \Re^p$ is a vector of measured outputs and $h : \Re^n \to \Re^p$ is a state to output mapping, and t is a continuous time variable. Many nonlinear control techniques are model-based in that they require the use of this type of dynamic model. A way of obtaining a nonlinear model of the plant is to derive it from physical principles. These models provide a rich insight into the process and are applicable over a wide range of operating conditions. However, physical models are often not available due to the high engineering effort and cost associated with their development and maintenance. Also, due to the complexity of industrial processes, physical models are unsuitable for control purposes. An alternative way for obtaining the required model is to identify it using measured input–output data. The development of techniques of system identification has made possible the synthesis of empirical dynamic models using data measured from the system [13]. In recent years, there has been considerable interest in developing nonlinear dynamic models from input–output data and a variety of model structures and techniques are available [10]. The predominant family of structures for obtaining nonlinear empirical models is known as artificial neural networks, which are inspired by the connectionism of biological neurons [14].

1.4 Neural Networks

Neural networks are distributed, adaptive and generally nonlinear learning machines built from many processing elements, which are often called neurons. Each processing element receives connections from other processing elements. The interconnectivity and number of processing elements define the network architecture. Neural networks are inspired by the connectionism of biological neurons and are capable of approximating arbitrary nonlinear input–output maps.

The field of neural computations has evolved from its neurological roots when the first artificial neural models were proposed [15], to its formalised mathematical foundations [16, 17, 14]. Neural networks can be divided into static and dynamic networks. A static neural model is described by an algebraic equation while a dynamic neural network is represented by a difference or differential equation depending on whether it is based on a discrete or continuous domain, respectively. The most common architectures for static neural networks are the single-layer feedforward networks [14], multilayer feedforward networks or multilayer perceptrons (MLPs) [16], [18], radial basis functions (RBF) [19, 20, 21, 22] and cerebellar model articulation controller (CMAC) networks [23, 24]. When feedback was introduced other relevant architectures were suggested, such as Hopfield networks [25, 26], Bolzmann machines [27, 28], Kohonen self-organising networks [29] and adaptive resonance networks [30].

The best-known type of neural network is the multilayer perceptron. The mathematics of the multilayer perceptron are briefly described below. Figure 1.3 shows a static processing element and associated signals.

6   1. Introduction

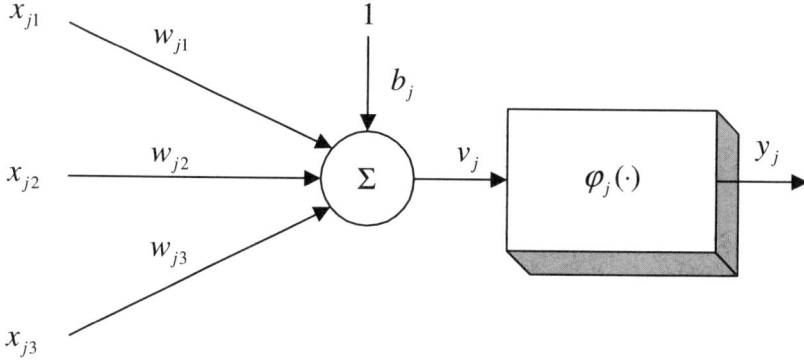

Fig. 1.3. Illustration of a static processing element or neuron

Define the inputs to a processing element j for a given sample n as u_{ji}, $i = 1, \cdots, m_j$. Then the input d_j to the activation of this processing element is given by:

$$d_j(n) = \sum_{i=1}^{m_j} w_{ji} u_{ji}(n) + b_j \qquad (1.8)$$

where w_{ji} the weight at the connection between input i and processing element j and b_j is the bias applied to processing element j.

The output of the processing element is the result of passing the scalar value $d_j(n)$ through its activation function $\sigma_j(\cdot)$:

$$z_j(n) = \sigma_j(d_j(n)) \qquad (1.9)$$

The actual shape of the activation function σ_j varies between applications. Figure 1.4 shows two different activation functions: the linear activation function and the hyperbolic tangent activation function.

Figure 1.5 shows a typical three-layer perceptron, with three input signals in the input layer, one hidden layer with two nodes and a single output signal.

For the multilayer perceptron shown in Figure 1.5 it is possible to write the output variable as follows:

$$z = \sigma_2(W_2\, \sigma_1(W_1 u + b_1) + b_2) \qquad (1.10)$$

where $u \in \Re^3$ is an input vector, $\sigma_1 : \Re^2 \to \Re^2$ is the activation function of the hidden layer, $\sigma_2 : \Re \to \Re$ is the activation function of the output layer, $b_1 \in \Re^2$ is a vector associated with the hidden layer, b_2 is the scalar bias associated with the output layer, $W_1 \in \Re^{2 \times 3}$ is a weight matrix associated with the hidden layer, $W_2 \in \Re^{1 \times 2}$ is a weight matrix associated with the output layer.

A training algorithm is used to adjust the weights of the interconnections according to the training data, which consists of input and output values that

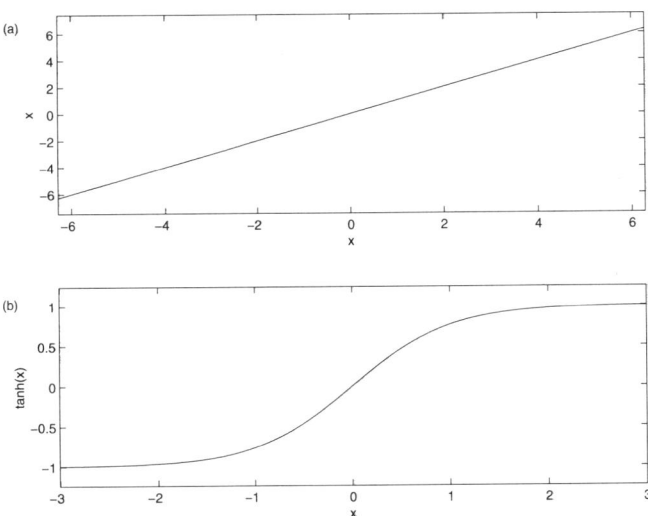

Fig. 1.4. Two common activation functions. (a) Linear $\sigma(x) = x$; (b) hyperbolic tangent $\sigma(x) = \tanh(x)$

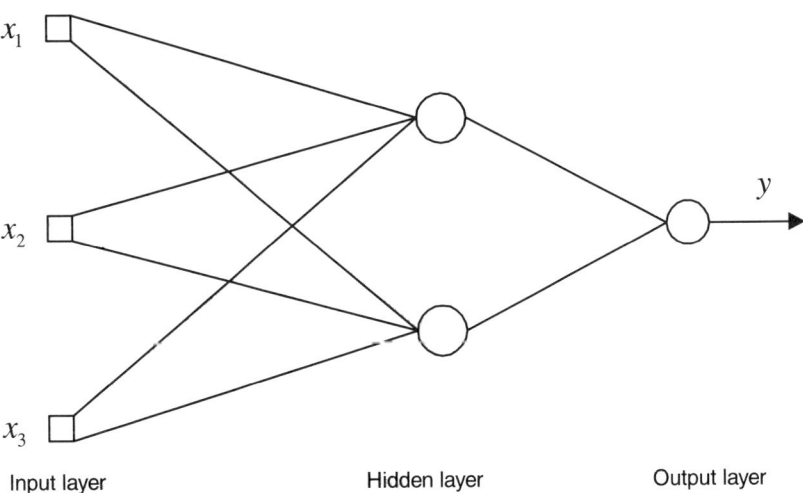

Fig. 1.5. A typical multilayer perceptron

the network is required to learn. The most widely used training methods to determine the weights of a multilayer perceptron is known as the *backpropagation algorithm* [14]. There are several variants of the backpropagation algorithm.

1.5 System Identification

System identification is the experimental approach to modelling dynamical systems. Using system identification techniques, it is possible to find dynamic models of systems based on measured input–output data. System identification involves the following steps [31, 32]:

- **Experiment.** The data needed to identify dynamic models is collected from the system. This involves changing the inputs in an appropriate way to excite the system, and measuring the output history over a period of time, as illustrated in Figure 1.6. The result of an identification experiment is a set of M discrete data points, $Z_M = \{\,[y(t_k),\,u(t_k)],\ k=1,\cdots,M\,\}$, where y and u are vectors of measured and input variables, respectively. The resulting data set may need some pre-processing (e.g. filtering) before it can be used for parameter estimation.
- **Model structure selection.** A model structure is a set of possible models with a number of free parameters. The choice of model structure is dependent on the purpose of the model, but the designer often has to choose between linear and nonlinear modes, input–output or state space descriptions, continuous or discrete-time models, etc.
- **Parameter estimation.** The free parameters of a model are estimated using an optimisation procedure that is typically aimed at minimising the differences between the measured outputs and the model outputs. Depending on the model structure used, simple non-iterative techniques like the least squares method may be used, although for nonlinear model structures iterative procedures are required.
- **Model validation.** This step involves evaluating the model to see if it satisfies the requirements for acceptance, which are in turn connected to the purpose of the model. A data set different from the one used for parameter estimation is typically employed for validating the model. If the model is not acceptable, then it may be necessary to repeat some of the above steps.

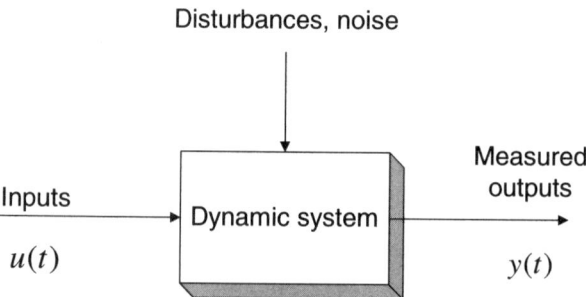

Fig. 1.6. The system identification experiment

1.6 Static Neural Networks for Identification and Control

Given the ability of multilayer perceptrons to approximate arbitrary continuous functions, neural networks have found wide applications both for function approximation and pattern recognition. Since the seminal work by Narendra and Parthasarathy [33], in which the use of neural networks was proposed for identification and control, a great deal of progress has been made in this field. The purpose of this section is to introduce a number of basic ideas on the use of static neural networks for identification and control.

General nonlinear discrete-time input–output models can be described as follows[1]:

$$y(t) = \hat{y}(t|\theta) + e(t) = g(\sigma(t,\theta), \theta) + e(t) \qquad (1.11)$$

where $y \in \Re^p$ is the output of the system, $\hat{y} \in \Re^p$ is the output of the model, $g : \Re^{n_\sigma} \times \Re^{n_\theta} \to \Re^p$ is a nonlinear mapping, $\sigma \in \Re^{n_\sigma}$ is known as the regression vector (which depends on past input–output information), $\theta \in \Re^{n_\theta}$ is the parameter vector, and $e \in \Re^p$ is known as the model residual.

A common nonlinear discrete-time structure, which is illustrated in Figure 1.7, is known as nonlinear auto-regressive with exogenous input (NARX). For this structure, the regression vector is given by:

$$\sigma(t,\theta) = [y(t-1), \cdots, y(t-n_a), u(t-d), \cdots, u(t-d-n_b+1)]^T \qquad (1.12)$$

where n_a is the number of past outputs, n_b is the number of past inputs and d an integer representing pure delay. The values of n_a, n_b, and d determine the size of the regression vector and the external structure of the model.

A multilayer perceptron is often used with the NARX structure to provide the static mapping g between the model input (the regression vector $\sigma(t,\theta)$) and the model output $\hat{y}(t|\theta)$. In that case, there is an internal structural choice in terms of the number of hidden layers and the number of neurons in each hidden layer. Once the structure is fixed, the vector of free parameters θ contains all the weights and biases associated with the multilayer perceptron. In this way, a static neural network can be used for modelling a discrete-time dynamical system.

Many strategies have been proposed to use neural networks for control. A recent review was provided by Agarwal [34]. Neural network based control may be classified as being *direct* or *indirect*, depending on the role played by neural networks in the control strategy. Direct neural network based control implies the use of a neural network as the controller. Indirect neural network based control involves the use of a neural network as an aid for modelling, control action or supervisory action.

---

[1] Notice that in the case of discrete-time models like Equation 1.11, variable t represents an integer time index.

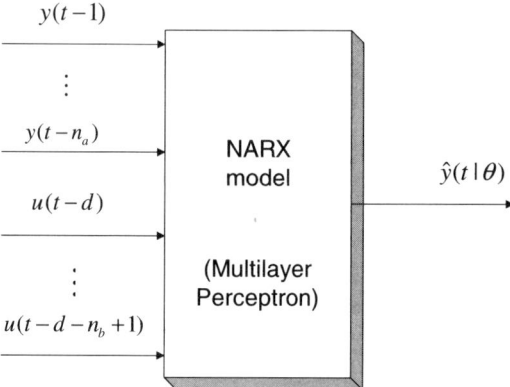

Fig. 1.7. The NARX structure

Direct inverse control. One of the first and simplest strategies proposed for direct neural network based control was the use of a neural network to approximate the inverse mapping of a system. Assume for simplicity that the system is SISO. If the controlled system can be described by:

$$y(t+1) = g(y(t), \cdots, y(t-n_a+1), u(t), \cdots, u(t-n_b+1)) \qquad (1.13)$$

then, the inverse model of the system computes the required input $u(t)$ to achieve a desired output $y^{(d)}(t+1)$ at the next time step:

$$u(t) = \hat{g}^{-1}(y^{(d)}(t+1), y(t), \cdots, y(t-n_a+1), u(t), \cdots, u(t-n_b+1)) \qquad (1.14)$$

where the mapping $\hat{g}^{-1}$ may be obtained by training a static neural network, such as a multilayer perceptron [35]. This approach is illustrated in Figure 1.8. Direct inverse control is simple and intuitive, but problems occur when the inverse models are not well damped or unstable. Also, this approach is very sensitive to noise and disturbances.

Internal model control (IMC). In this direct strategy, the difference e between the output of a neural network forward model $\hat{y}$ and the output of the plant y is used as feedback signal, while a neural network approximating the inverse of the system is placed in the forward path [36]. If the forward model is perfect, the error signal will be zero and the control system will operate as if it was under direct inverse control. This approach is illustrated in Figure 1.9. The forward model can be a multilayer perceptron based NARX model.

1.7 Dynamic Neural Networks for Nonlinear Identification

The introduction of feedback into a feedforward neural network architecture produces a state space dynamic model. A dynamic recurrent neural network

1.7 Dynamic Neural Networks for Nonlinear Identification

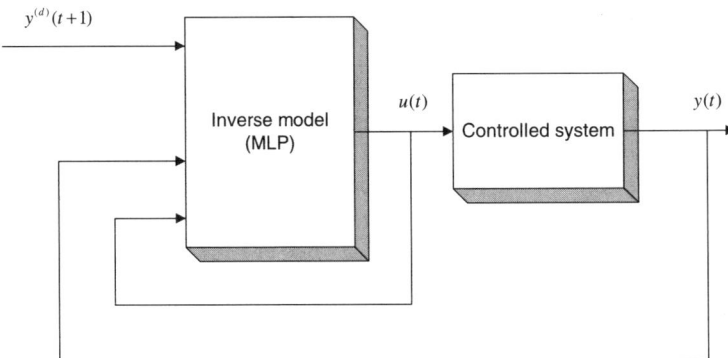

Fig. 1.8. Direct inverse control

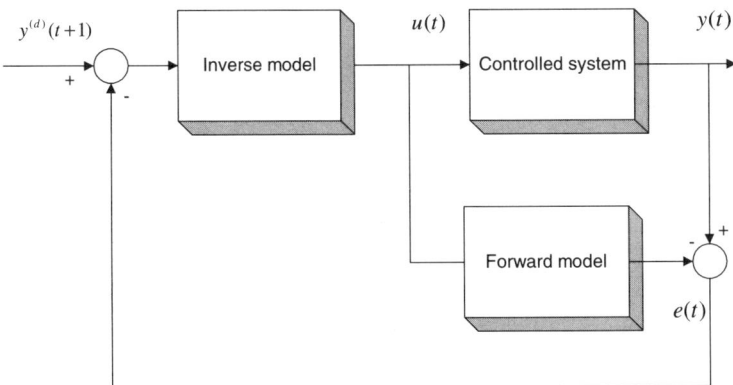

Fig. 1.9. Internal model control

(DRNN), or simply a dynamic neural network (DNN), is a collection of dynamic neurons partially interconnected to a function of their own output. Such networks can be represented by a state space neural model of the form

$$\dot{x} = -\beta x + \omega \sigma(x) + \gamma u \qquad (1.15)$$
$$y_n = C_n x$$

where x are coordinates on $\Re^N$, $\omega \in \Re^{N \times N}$, $\sigma(x) = [\sigma(x_1), \ldots, \sigma(x_N)]^T$, $\gamma \in \Re^{N \times p}$, $u \in \Re^p$, $C_n = [I_{p \times p} 0_{p \times (N-p)}]$ and $\beta \in \Re^{N \times N}$ is a diagonal matrix with diagonal elements $\{\beta_1, \ldots, \beta_N\}$.

The use of a dynamic neural network model for system identification purposes seems now a straightforward task. Introducing inherent dynamic capabilities in the neurons has made possible that an array of these elements be used as generic model for system identification.

1.8 Input–Output Linearisation-Decoupling and the Use of Dynamic Neural Networks

Many industrial processes are multivariable in the sense that they have several inputs and outputs. In addition to nonlinear behaviour, such processes often exhibit interactions, in the sense that a change in an input variable may originate changes in several output variables. In a multivariable system, in addition to linearising the behaviour from inputs to outputs, it is possible to use feedback in order to eliminate or weaken the interactions and reduce the system, at least from an input–output perspective, to several independent single-input single-output systems. This synthesis method is known as the input–output linearisation-decoupling scheme and it requires a model of the system. Instead of using a process model derived from physical considerations, the control strategies proposed in this book are based on dynamic neural network models as given in Equation 1.15. These models approximate the behaviour of the system and are trained using input–output data. A feedback linearising-decoupling law is synthesised for the neural model and applied on the MIMO plant. Once the dynamics are linearised and decoupled to a certain extent, the system is immersed in an outer loop with a standard multivariable Proportional+Integral (PI) scheme. The use of neural networks for feedback linearisation started recently and similar general motivations appear to underlie recent approaches where the existence of a neural network model encouraged the use of feedback linearising techniques. Feedback linearisation of SISO systems has been carried out by means of feedforward networks [37, 38], while other work has used dynamic networks from a robust control perspective [39]. Input–output linearisation of multivariable processes within an inverse model control framework has also been carried out [40, 41]. Also for SISO systems, input–output linearisation techniques have been proposed using dynamic neural networks [42] and CMAC neural networks [43]. Later work suggested the use of discrete time neural models for feedback linearisation of chemical processes [44]. In recent work, an adaptive linearising feedback technique has been developed for induction motor control based on dynamic neural networks [45].

1.9 Potential applications

Model predictive control (MPC) is a technique that periodically uses a model of the controlled system to calculate the control action based on optimal input–output predictions over a time horizon. This technique has enjoyed remarkable industrial success since the first reported industrial implementations in the late 1970s [46]. Key features contributing to its success are that multivariable systems and constraints can be accommodated effectively in the control problem, and the use of empirical models which can be built from measured input–output data. The traditional application areas were refining and petrochemicals, but significant growth in areas such as chemicals, pulp and paper, food processing, aerospace and automotive industries has been noticed in the last few years [47]. To date, most commercial predictive controllers employ linear empirical models. However, nonlinear empirical models are starting to be used in some commercial predictive control packages [48]. The resulting approach is known as nonlinear predictive control. The control problem is analogous to the linear version of MPC except that a nonlinear dynamic model is used. It can be expressed as determining the control actions by solving a nonlinear programming problem at each sampling interval in order to minimise the error between the outputs and their setpoints over the optimisation horizon [49, 50]. One of the drawbacks of this approach is the great computational burden associated with the on–line solution of the nonlinear programming problem [51]. Feedback linearisation techniques based on empirical nonlinear models can potentially be used to enable MPC techniques to consider nonlinearities, while at the same time have the computational advantage of a linear model, as preliminary research has shown [52, 53, 54, 55].

CHAPTER 2
FUNDAMENTAL CONCEPTS

In order to understand many of the topics presented in later chapters of this book, it is important to introduce and keep in mind a number of concepts of geometric theory. Section 2.1 serves this purpose. An understanding of the stability theory for nonlinear systems is also relevant for the purposes of this book, so the fundamentals of nonlinear stability theory based on the approach by Lyapunov are presented in Section 2.2.

2.1 Elementary Concepts of Geometric Theory

2.1.1 Linear vector spaces

Definition 2.1.1 (Linear vector space). *A linear vector space is a set V whose elements are known as vectors, and two operations known as addition and multiplication, such that: (i) For given vectors $x, y, z \in V$, the addition is defined $x + y \in V$, $x + y = y + x$, $(x + y) + z = x + (y + z)$. (ii) There is a zero vector $0 \in V$, such that for all $x \in V$, $x + 0 = x$. (iii) For any real numbers r_1 and r_2 and any vectors $x, y \in V$, the multiplication is defined, $r_1 x \in V$, $1.x = x$, $0.x = 0$, $r_1(r_2 x) = (r_1 r_2)x$, $r_1(x + y) = r_1 x + r_1 y$, $(r_1 + r_2)x = r_1 x + r_2 x$.*

Definition 2.1.2 (Subspace). *A subset W of a linear vector space is called a subspace of V if W satisfies the following conditions: (i) If $x, y \in W$, then $x + y \in W$. (ii) If $x \in W$ and $r \in \Re$, then $rx \in W$.*

2.1.2 Euclidean space

The set of all possible n-dimensional vectors $x = [x_1, \ldots, x_n]^T$, where x_j, $j = 1, \ldots, n$, are real numbers, defines the Euclidean space denoted by $\Re^n$. Vectors in $\Re^n$ can be added by adding their individual components. Given $x = [x_1, \ldots, x_n]^T \in \Re^n$ and, $y = [y_1, \ldots, y_n]^T \in \Re^n$, then:

$$x + y = [x_1 + y_1, \ldots, x_n + y_n]^T \tag{2.1}$$

Furthermore, a real number α can be multiplied by a vector $x = [x_1, \ldots, x_n]^T$ as follows:

$$\alpha x = [\alpha x_1, \ldots, \alpha x_n]^T \qquad (2.2)$$

With these two operations, the Euclidean space is a linear vector space.

2.1.3 Vector norms

The norm of a vector $x = [x_1, \ldots, x_n]^T$, denoted as $||x||$, is a real valued function that has the following properties:

- $||x|| = 0$ if and only if $x_j = 0, j = 1, \ldots, n$, which is denoted as $x = 0$.
- $||x|| > 0$ for all $x \in \Re^n$, $x \neq 0$.
- $||\alpha x|| = |\alpha| ||x||$, for all $x \in \Re^n$ and $\alpha \in \Re$.
- $||x + y|| \leq ||x|| + ||y||$ for all $x, y \in \Re^n$.

In this book, we will consider the family of p-norms defined by:

$$||x||_p = (|x_1|^p + \cdots + |x_n|^p)^{1/p} \qquad (2.3)$$

where $1 \leq p < \infty$. The well known Euclidian norm is a p-norm with $p = 2$, and is denoted as $||x||_2$. The infinity norm is defined as follows:

$$||x||_\infty = \max_i |x_i| \qquad (2.4)$$

2.1.4 Matrix norms

A matrix A with n rows and m columns and real elements, $a_{ij} \in \Re, i = 1, \ldots, n, j = 1, \ldots m$, defines a linear mapping $x = Az$ from $\Re^m$ to $\Re^n$. The induced p-norm of matrix A is defined as follows:

$$||A||_p = \sup_{x \neq 0} \frac{||Az||_p}{||z||_p} = \max_{||z||_p = 1} ||Az||_p \qquad (2.5)$$

where sup stands for supremum or least upper bound. In particular, for $p = 2$ and $p = \infty$, we have:

$$||A||_2 = \left[\max\{\text{eig}(A^T A)\}\right]^{1/2} \qquad (2.6)$$

$$||A||_\infty = \max_i \sum_{j=1}^n |a_{ij}| \qquad (2.7)$$

so that the induced 2-norm is the square root of the maximum eigenvalue of $A^T A$ (or maximum singular value), and the induced ∞-norm is the maximum row sum of absolute element values.

Induced matrix norms satisfy the following properties:

$$||Ax|| \leq ||A|| \, ||x|| \qquad (2.8)$$

$$||A + B|| \leq ||A|| + ||B|| \qquad (2.9)$$

$$||AB|| \leq ||A|| \, ||B|| \qquad (2.10)$$

where A and B are matrices of appropriate dimensions, and x is a vector.

2.1.5 Sets

Definition 2.1.3 (Open and closed sets). *A subset $U \subset \Re^n$ is said to be open if for every vector $x \in U$, a neighbourhood of x can be found,*

$$N(x, \delta) = \{y \in \Re^n \mid ||y - x|| < \delta\} \tag{2.11}$$

such that $N(x, \delta) \subset U$. On the other hand, a set W is closed if and only if its complement in $\Re^n$ is an open set.

Definition 2.1.4 (Boundary of a set). *A point x is said to be a boundary point of a subset $U \subset \Re^n$ if every neighbourhood of x contains at least one point of U and one point not belonging to U. The set of all boundary points of U, which is denoted by ∂U, is called the boundary of U.*

Definition 2.1.5 (Interior and closure of a set). *The interior of a set S is $S - \partial S$. The closure of a set S, denoted by $\bar{S}$, is the union of S and its boundary. An open set is equal to its interior. A closed set is equal to its closure.*

Definition 2.1.6 (Bounded set). *A subset $U \subset \Re^n$ is bounded if there is a $r > 0$ such that $||x|| < r$ for all $||x|| \in U$.*

Definition 2.1.7 (Compact set). *A set $S \subset \Re^n$ is compact if it is closed and bounded.*

Definition 2.1.8 (Convex set). *A subset $U \subset \Re^n$ is convex if for every $x, y \in U$ and every real r, with $0 < r < 1$, the point $z = rx + (1 - r)y \in U$.*

2.1.6 Vector fields

The definitions of vector field, covector field and their differential operators are important for the understanding of differential geometry approaches for the control of dynamic systems [5, 9].

Definition 2.1.9 (Vector field). *A vector field F on an open subset U of $\Re^n$ is a function that assigns to each $x \in U$ a column vector $F(x) \in \Re^n$. In short, $F : U \to \Re^n$, $U \subset \Re^n$.*

A vector field F on $\Re^n$ will be considered smooth or differentiable if all partial derivatives $\partial F_i / \partial x_j$ of arbitrary order exist and are continuous functions of $(x_1, \ldots, x_n)$. Together with vector fields, it is convenient to take into account dual objects called covector fields.

Definition 2.1.10 (Covector field). *The transpose of a vector field is known as a covector field.*

2. Fundamentals concepts

A covector field of special importance is the partial derivative of a real valued differentiable function λ defined on an open subset of $\Re^n$,

$$\frac{\partial \lambda}{\partial x} = \left[\frac{\partial \lambda}{\partial x_1}, \ldots, \frac{\partial \lambda}{\partial x_n} \right] \tag{2.12}$$

A key concept in differential geometry is the notion of how to differentiate a vector field. For vector fields on open subsets, the definition is natural.

2.1.7 Differential operations

There are two main differential operations involving vector fields and covector fields that are used in this book: (i) The derivative of a function along a vector field or Lie derivative and (ii) the Lie product.

Definition 2.1.11 (Lie derivative). *Let $\lambda : \Re^n \to \Re$ be a differentiable function and a vector field f, both defined on an open subset U of $\Re^n$. The derivative of λ along f or Lie derivative of λ along f is given by the inner product*

$$\left\langle \frac{\partial \lambda}{\partial x}, f(x) \right\rangle = \frac{\partial \lambda}{\partial x} f(x) \tag{2.13}$$

The Lie derivative of λ along f is usually denoted as $L_f \lambda$, so that:

$$L_f \lambda(x) = \sum_{i=1}^{n} \frac{\partial \lambda}{\partial x_i} f_i(x) \tag{2.14}$$

for each given $x \in U$.

If $L_f \lambda$ is again differentiated along another vector field g, the following is obtained:

$$L_g L_f \lambda(x) = \frac{\partial (L_f \lambda)}{\partial x} g(x) \tag{2.15}$$

This operation could be used recursively along the same vector field. $L_f^k \lambda$ indicates that λ is being differentiated k times along f such that

$$L_f^k \lambda(x) = \frac{\partial \left(L_f^{k-1} \lambda \right)}{\partial x} f(x) \tag{2.16a}$$

$$L_f^0 \lambda(x) = \lambda(x) \tag{2.16b}$$

Example 2.1.1. Suppose that $x = [x_1, x_2]^T$, $\lambda(x) = \sqrt{x_1}$ and $f(x) = [x_2, \sin x_1]^T$. Therefore the Lie derivative is:

$$L_f \lambda(x) = \frac{\partial \lambda}{\partial x} f(x) = \left[\frac{1}{2\sqrt{x_1}}, 0 \right] \begin{bmatrix} x_2 \\ \sin x_1 \end{bmatrix} = \frac{x_2}{2\sqrt{x_1}} \tag{2.17}$$

2.1 Elementary Concepts of Geometric Theory 19

Definition 2.1.12 (Lie product). *Let f and g be two vector fields, both defined on an open subset U of $\Re^n$. The Lie product or Lie bracket of f and g is a new differentiable vector field defined as:*

$$[f,g](x) = \frac{\partial g}{\partial x} f(x) - \frac{\partial f}{\partial x} g(x) \tag{2.18}$$

where the Jacobian matrices of the mappings g and f are given respectively by

$$\frac{\partial g}{\partial x} = \begin{bmatrix} \frac{\partial g_1}{\partial x_1} & \cdots & \frac{\partial g_1}{\partial x_n} \\ \vdots & \ddots & \vdots \\ \frac{\partial g_n}{\partial x_1} & \cdots & \frac{\partial g_n}{\partial x_n} \end{bmatrix}, \quad \frac{\partial f}{\partial x} = \begin{bmatrix} \frac{\partial f_1}{\partial x_1} & \cdots & \frac{\partial f_1}{\partial x_n} \\ \vdots & \ddots & \vdots \\ \frac{\partial f_n}{\partial x_1} & \cdots & \frac{\partial f_n}{\partial x_n} \end{bmatrix} \tag{2.19}$$

for each given $x \in U$.

Some important properties of the differential operations defined above are summarised in the following proposition. The proof is relatively straightforward [5].

Proposition 2.1.1. *The differential operators in Definitions 2.1.11 and 2.1.12 are characterised by the following properties:*

- *The Lie product of vector fields is bilinear over $\Re$. In other words, if f_1, f_2, g_1, g_2 are vector fields and r_1, r_2 real numbers then*

$$[r_1 f_1 + r_2 f_2, g_1] = r_1[f_1, g_1] + r_2[f_2, g_1] \tag{2.20a}$$
$$[f_1, r_1 g_1 + r_2 g_2] = r_1[f_1, g_1] + r_2[f_1, g_2] \tag{2.20b}$$

- *The Lie product of vector fields is skew commutative:*

$$[f, g] = -[g, f] \tag{2.21}$$

- *If α and λ are real-valued functions and f is a vector field, then*

$$L_{\alpha f} \lambda(x) = (L_f \lambda(x)) \alpha(x) \tag{2.22}$$

- *If α and β are real-valued functions and f and g are vector fields, then*

$$[\alpha f, \beta g](x) = \alpha(x)\beta(x)[f,g](x) + (L_f \beta(x))\alpha(x) g(x) \\ - (L_g \alpha(x))\beta(x) f(x) \tag{2.23}$$

- *If f, g are vector fields and λ a real-valued function, then*

$$L_{[f,g]} \lambda(x) = L_f L_g \lambda(x) - L_g L_f \lambda(x) \tag{2.24}$$

- *If f is a vector field and λ a real-valued function, then*

$$L_f \frac{\partial \lambda(x)}{\partial x} = \frac{\partial}{\partial x}[L_f \lambda(x)] \tag{2.25}$$

20   2. Fundamentals concepts

2.1.8 Distributions

Definition 2.1.13 (Distribution). *A distribution D is the vector space generated from the span of a group of vectors $f_i(x), i = 1, \ldots, q$ on an open set U of $\Re^n$:*

$$D(x) = \text{span}\{f_1(x), \ldots, f_q(x)\}$$

Notice that to each point $x \in U$ we assign a subspace $D(x)$ of $\Re^n$. In other words, D is a collection of all vector spaces $D(x)$ for $x \in U$.

Notice that $\dim(D) = \text{rank}\,[f_1(x), f_2(x), \ldots, f_q(x)]$, may vary with x. A distribution D is said to be *nonsingular* if $\dim(D(x)) = q$ for all $x \in U$.

Definition 2.1.14 (Involutive distribution). *A distribution D is said to be involutive if for all vector fields $g_1, g_2 \in D$ the Lie product $[g_1, g_2] \in D$.*

Example 2.1.2. Consider a distribution on $\Re^3$ defined as follows:

$$D = \text{span}\{f_1, f_2\} \tag{2.26}$$

where

$$f_1(x) = \begin{bmatrix} 5x_3 \\ -2 \\ 0 \end{bmatrix} \quad f_2(x) = \begin{bmatrix} -3x_1 \\ -6x_2 \\ 2x_3 \end{bmatrix} \tag{2.27}$$

Given that $f_1(x)$ and $f_2(x)$ are linearly independent for all $x \in \Re^3$, then the dimension of this distribution is $\dim(D(x)) = 2$ for all $x \in \Re^3$. Thus the distribution is nonsingular. To check if the distribution is involutive, we need to calculate the Lie product:

$$[f_1, f_2](x) = \frac{\partial f_2}{\partial x} f_1(x) - \frac{\partial f_1}{\partial x} f_2(x) \tag{2.28}$$

$$\tag{2.29}$$

$$= \begin{bmatrix} -3 & 0 & 0 \\ 0 & -6 & 0 \\ 0 & 0 & 2 \end{bmatrix} \begin{bmatrix} 5x_3 \\ -2 \\ 0 \end{bmatrix} - \begin{bmatrix} 0 & 0 & 5 \\ 0 & 0 & 0 \\ 0 & 0 & 0 \end{bmatrix} \begin{bmatrix} -3x_1 \\ -6x_2 \\ 2x_3 \end{bmatrix} = \begin{bmatrix} -25x_3 \\ 12 \\ 0 \end{bmatrix} \tag{2.30}$$

We then need to check if $[f_2, f_2](x) \in D$. To do this, we check the rank of a matrix with columns $f_1(x)$, $f_2(x)$ and $[f_1, f_2](x)$:

$$\text{rank}\left(\begin{bmatrix} 5x_3 & -3x_1 & -25x_3 \\ -2 & -6x_2 & 12 \\ 0 & 2x_3 & 0 \end{bmatrix}\right) = 2 \tag{2.31}$$

As this matrix has not full rank, then we conclude that $[f_1, f_2](x)$ is not linearly independent from f_1 and f_2, so that we can generate $[f_1, f_2](x)$ as a linear combination of $f_1(x)$ and $f_2(x)$. Therefore, $[f_1, f_2](x) \in D$ and the distribution D is involutive. ∎

Definition 2.1.15 (Complete Integrability). *Let Δ be a non-singular distribution on D, generated by $f_1, \ldots, f_r$. Then, Δ is said to be completely integrable if for each $x_o \in D$, there is a neighbourhood N of x_o and $n - r$ functions $h_1(x), \ldots, h_{n-r}(x)$ that satisfy the differential equations*

$$\frac{\partial h_j}{\partial x} f_i(x) = 0, \quad \text{for all } i = 1, \ldots, r \text{ and } j = 1, \ldots, n - r \qquad (2.32)$$

and the covector fields $\partial h_j(x)/\partial x$ are linearly independent for all $x \in D$.

Theorem 2.1.1 (Frobenius). *A nonsingular distribution is completely integrable if and only if it is involutive.*

Proof. See [5].

Theorem 2.1.1 is very important in the analysis of nonlinear control systems using the differential-geometric approach, as it establishes a correspondence between the notion of involutive distribution (a property that is easy to verify) and the concept of complete integrability, which is fundamental to define the normal form of nonlinear systems, as will be seen in Chapter 3.

2.2 Stability of Nonlinear Systems

Stability is a very important issue in the study of nonlinear systems and an understanding of the fundamental principles of nonlinear stability is required later in this book. This section provides an introduction to the main definitions and theorems associated with Lyapunov stability. Further details and examples can be found, for instance, in the book by Khalil [56].

Definition 2.2.1 (Lipschitz condition). *A vector field $g(x)$ is said to satisfy a Lipschitz condition on an open set U of $\Re^n$ if there is a constant L, such that*

$$\|g(x_1) - g(x_2)\| \leq L\|x_1 - x_2\| \qquad (2.33)$$

for all $x_1, x_2 \in U$. $g(x)$ is then said to be Lipschitz continuous (or just Lipschitz) on U and L is known as a Lipschitz constant. The Lipschitz constant usually is taken as the least value of L that satisfies the inequality.

If $g(x)$ is Lipschitz in x, then it is continuous in x. If g is continuous and has bounded partial derivatives in x, then it is Lipschitz. If the following relationship holds:

$$\left\|\frac{\partial g}{\partial x}\right\| \leq L \qquad (2.34)$$

then $g(x)$ is Lipschitz with constant L [56].

2. Fundamentals concepts

Example 2.2.1. Consider the vector field:

$$g(x) = Ax \tag{2.35}$$

where $x \in \Re^n$ and $A \in \Re^{n \times m}$. Consider also the vectors x and $y \in \Re^n$. We have that:

$$||g(x) - g(y)||_p = ||Ax - Ay||_p = ||A(x-y)||_p \leq ||A||_p ||x-y||_p \tag{2.36}$$

where the last inequality is a consequence of Equation 2.8. Then we conclude that a Lipschitz constant for $g(x) = Ax$ is the induced norm $L = ||A||_p$. Notice that the value of the Lipschitz constant depends on the norm used. ∎

Consider a nonlinear system described by:

$$\dot{x} = f(x,t); \quad x(t_0) = x_0 \tag{2.37}$$

where $x \in U \subset \Re^n$, and $f : U \times \Re \to \Re^n$ is a vector field.

Definition 2.2.2 (Equilibrium point). *A point $x^* \in U \subset \Re^n$ is called an equilibrium point of a system described by Equation 2.37, if $f(x^*, t) = 0$ for all $t \geq 0$.*

It is assumed below that the system in Equation 2.37 has solutions $x(t)$ for $t > t_0$, in a neighbourhood U of an equilibrium point, which is guaranteed if f is piecewise continuous in t and locally Lipschitz in x in a neighbourhood of an equilibrium point x^*.

If we translate the origin to an equilibrium point $x^* \in U \subset \Re^n$, then we can make the origin $x = 0$ an equilibrium point in the new coordinates. This assumption made below and has great notational utility.

Definition 2.2.3 (Stability in the sense of Lyapunov). *An equilibrium point $x = 0 \in U \subset \Re^n$ of a system described by Equation 2.37 is called stable if for all $t_0 \geq 0$ and $\epsilon > 0$, there exists a $\delta(t_0, \epsilon)$, such that*

$$||x_0|| < \delta(t_0, \epsilon) \Rightarrow ||x(t)|| < \epsilon, \quad \text{for all} \quad t \geq t_0 \tag{2.38}$$

where $x(t)$ is the solution of Equation 2.37 from the initial condition $x(t_0) = x_0$. If δ can be chosen independent of t_0 then $x = 0$ is said to be uniformly stable.

Example 2.2.2. The linear time-varying system described by:

$$\dot{x}(t) = \cos(t)x(t), \quad x(t_0) = x_0 \tag{2.39}$$

has the origin $x = 0$ as an equilibrium point and has the solution:

$$x(t) = x_0 \exp\left(-\sin(t_0) + \sin(t)\right) \tag{2.40}$$

2.2 Stability of Nonlinear Systems

For any value of t_0 it is obvious that the exponential argument is bounded, so that the exponential term is bounded by a constant $c(t_0)$ that depends on t_0, so that:

$$|x(t)| < |x(t_0)|c(t_0), \quad t > t_0 \tag{2.41}$$

For any $\epsilon > 0$ if we choose $\delta(t_0) = \epsilon/c(t_0)$ such that for $x(t_0) < \delta(t_0)$, then

$$|x(t)| < |x(t_0)|c(t_0) < \delta(t_0)c(t_0) = \epsilon \tag{2.42}$$

which shows that the origin $x = 0$ is stable. However, since $\delta(t_0)$ cannot be chosen independent of t_0, then the origin is not uniformly stable. ∎

Definition 2.2.4 (Asymptotic stability). *Assume that $x = 0$ is an equilibrium point of a system described by Equation 2.37. Then $x = 0$ is said to be asymptotically stable if it is stable and attractive, that is, for all $t_0 \geq 0$ there is a $\delta(t_0)$ such that*

$$||x_0|| < \delta \Rightarrow \lim_{t \to \infty} ||x(t)|| = 0 \tag{2.43}$$

If δ can be chosen independent of t_0 then $x = 0$ is said to be uniformly asymptotically stable. *If $\delta \to +\infty$ then $x = 0$ is said to be* globally asymptotically stable.

Definition 2.2.5 (Region of attraction). *Suppose that $x = 0$ is an attractive equilibrium point of a system described by Equation 2.37. The region of attraction $R(0)$ is defined as the set of initial states $x(0) = x_0$ for which the resulting trajectory $x(t)$, $t \geq 0$, tends to the equilibrium point:*

$$R(0) = \{ \ x_0 \in \Re^n | x(t) \to 0 \text{ as } t \to \infty \} \tag{2.44}$$

Definition 2.2.6 (Exponential stability). *Let $x = 0$ be an equilibrium point of a system described by Equation 2.37 and $U \subset \Re^n$ a domain containing it. x is said to be exponentially stable if there exist scalars $m > 0$ and $c > 0$ such that the solution $x(t)$ satisfies:*

$$||x(t)|| \leq m||x_0|| \exp(-c(t - t_0)), \text{ for all } x_0 \in U \text{ and } t \geq t_0 \tag{2.45}$$

The constant c is called the rate of convergence.

The stability definitions given above are concerned with the solutions of differential equations in a neighbourhood of an equilibrium point. Intuitively, stability implies that the solution $x(t)$ stays near the equilibrium point, while asymptotic stability implies that the solution tends to the equilibrium point as $t \to \infty$. Uniform asymptotic stability means that the convergence of the solution to the equilibrium point is independent of the initial time t_0. Exponential stability of an equilibrium point always implies uniform asymptotic stability, while the converse is not true.

Definition 2.2.7 (Class K functions). *A scalar function* $\alpha : \Re_+ \to \Re_+$ *belongs to class K, if it is continuous, strictly increasing and* $\alpha(0) = 0$. *This is denoted as* $\alpha \in K$. *The function* α *is said to belong to class* K_∞ *if* $\alpha(r) \to \infty$ *as* $r \to \infty$.

Definition 2.2.8 (Positive definite functions). *A continuously differentiable function* $W : \Re^n \to \Re_+$ *is said to be positive definite in a region U of* $\Re^n$ *that contains the origin if (1)* $W(0) = 0$ *and (2)* $W(x) > 0$ *for* $x \in U$ *and* $x \neq 0$. $W(x)$ *is said to be positive semidefinite if* $W(x) >= 0$ *for* $x \in U$ *and* $x \neq 0$.

Conversely, if condition (2) in Definition 2.2.8 is replaced by $W(x) < 0$, then $W(x)$ is said to be *negative definite*. $W(x)$ is said to be *negative semidefinite* if $W(x) \leq 0$.

Definition 2.2.9 (Decrescent functions). *A function* $V(x,t) : \Re^n \times \Re \to \Re$ *is called decrescent in* $x \in U \subset \Re^n$ *and* $t \in [t_0, \infty]$, *if there is a function* $\beta : \Re \to \Re$, $\beta \in K$, *such that*

$$V(x,t) \leq \beta(||x||), \quad \text{for all } x(t) \in U \text{ and } t \geq t_0 \tag{2.46}$$

Theorem 2.2.1 (Lyapunov uniform asymptotic stability of nonautonomous systems). *Let* $x = 0$ *be an equilibrium point of a system described by 2.37 and* $U \subset \Re^n$ *a domain containing it. Let* $V : [0, \infty] \times U \to \Re$ *be a continuously differentiable function that satisfies:*

$$W_1(x) \leq V(x,t) \leq W_2(x) \tag{2.47}$$

$$\dot{V}(x,t) = \frac{\partial V}{\partial t} + \frac{\partial V}{\partial x} f(x,t) \leq -W_3(x) \tag{2.48}$$

for all $t \geq t_0$, *and* $x \in U$, *where* $W_1(x)$, $W_2(x)$ *and* $W_3(x)$ *are continuous positive definite functions on U. Then,* $x = 0$ *is uniformly asymptotically stable and V is called a Lyapunov function.*

Proof. The proof of this Theorem can be found in [56].

Corollary 2.2.1. *Suppose that the assumptions of Theorem 2.2.1 hold for all* $x \in \Re^n$ *and* $W_1(x) \to \infty$ *for* $||x|| \to \infty$, *then* $x = 0$ *is globally uniformly asymptotically stable.*

The proof of this Corollary can be found in [56].

Example 2.2.3. Consider the scalar system described by:

$$\dot{x} = -x^3 + \frac{x^3}{2}\sin(t), \quad x(t_0) = x_0 \tag{2.49}$$

Using the Lyapunov function candidate,

$$V(x) = \frac{1}{2}x^2 \qquad (2.50)$$

we obtain its time derivative along the trajectories of the system:

$$\dot{V}(x) = \frac{\partial V}{\partial x}\left[-x^3 + \frac{x^3}{2}\sin(t)\right] = x\left[-x^3 + \frac{x^3}{2}\sin(t)\right] = -x^4\left[1 - \frac{\sin(t)}{2}\right] \qquad (2.51)$$

If we choose $W_1(x) = W_2(x) = V(x)$ and $W_3(x) = x^4$, then the assumptions of Theorem 2.2.1 are satisfied globally. Therefore, the origin $x = 0$ is globally uniformly asymptotically stable. ∎

Corollary 2.2.2. *Suppose that the assumptions of Theorem 2.2.1 are replaced by:*

$$c_1||x||^q \leq V(x,t) \leq c_2||x||^q \qquad (2.52)$$

$$\dot{V}(x,t) \leq -c_3||x||^q \qquad (2.53)$$

for some positive constants c_1, c_2, c_3 and q. Then $x = 0$ is exponentially stable. Furthermore, if the assumptions are satisfied for all $x \in \Re^n$, then $x = 0$ is globally exponentially stable.

Proof. The proof of this Corollary can be found in [56].

For the case of autonomous systems, the stability theorem of Lyapunov can be stated as follows:

Theorem 2.2.2 (Lyapunov stability of autonomous systems). *Let $x = 0$ be an equilibrium point for a system described by:*

$$\dot{x} = f(x) \qquad (2.54)$$

where $f : D \to \Re^n$ is a locally Lipschitz and $U \subset \Re^n$ a domain that contains the origin. Let $V : U \to \Re$ be a continuously differentiable, positive definite function in U.

1. *If $\dot{V}(x) = [\partial V/\partial x]f$ is negative semidefinite, then $x = 0$ is a stable equilibrium point.*
2. *If $\dot{V}(x)$ is negative semidefinite, then $x = 0$ is an asymptotically stable equilibrium point.*

In both cases above V is called a Lyapunov function. Moreover, if the conditions hold for all $x \in \Re^n$ and $||x|| \to \infty$ implies that $V(x) \to \infty$, then $x = 0$ is globally stable in case 1 and globally asymptotically stable in case 2.

Proof. The proof of this Theorem can be found in [56].

26   2. Fundamentals concepts

The Theorems and Corollaries of stability given above characterise the stability of an equilibrium point by requiring the existence of a Lyapunov function $V(x,t)$ that satisfies some conditions. One of the beauties of the Lyapunov approach is that the differential equations do not need to be solved to ascertain the stability of an equilibrium point. This theory, however, does not explain how to find suitable Lyapunov functions. In the case of some physical systems, a suitable Lyapunov function is the energy function, as Lyapunov functions can be interpreted as generalised energy functions. However, in many cases, a trial and error approach to finding a Lyapunov function is required. It is useful to know if such a function exists, so the question arises as to under what circumstances the existence of a Lyapunov function can be inferred assuming the stability behaviour to be known. The answer is provided by the converse Lyapunov theorems [57], one of which is presented below.

Theorem 2.2.3 (Converse Lyapunov theorem). *Assume that $f : U \times \Re \to \Re^n$, with $U = \{x \in \Re^n \mid ||x|| < h\}$ has continuous and bounded first order partial derivatives in x and is piecewise continuous in t for all $x \in U$, $t \geq t_0$. Then the following statements are equivalent:*

- *$x = 0$ is an exponentially stable equilibrium point of a system described by Equation 2.37.*
- *There exists a function $V(x,t)$ and some strictly positive constants h', c_1, c_2, c_3, c_4, such that for all $x \in U' = \{x \in \Re^n \mid ||x|| \leq h'\}$ and $t \geq t_0$:*

$$c_1 ||x||^2 \leq V(x,t) \leq c_2 ||x||^2 \tag{2.55}$$

$$\dot{V}(x,t) \leq -c_3 ||x||^2 \tag{2.56}$$

$$\left|\left|\frac{\partial V(x,t)}{\partial x}\right|\right| \leq c_4 ||x|| \tag{2.57}$$

Proof. The proof of this Theorem can be found in [58].

2.3 Summary

This chapter has introduced a number of concepts that will be useful for the rest of the book. These include basic concepts of geometric theory, Lie derivatives and products, and of stability analysis of nonlinear dynamic systems based on Lyapunov theory.

CHAPTER 3
INTRODUCTION TO FEEDBACK LINEARISATION

Feedback linearisation is perhaps the most important nonlinear control design strategy developed during the last few decades [5]. The main objective of the approach is to algebraically transform a nonlinear dynamic system into a linear dynamic system by using a static state feedback and a nonlinear coordinate transformation based on a differential geometric analysis of the system. By eliminating nonlinearities in the closed loop system, conventional linear control techniques can be applied. Our presentation of relative degree, normal forms, zero dynamics and feedback linearisation for nonlinear control affine systems given in Section 3.1 follows the notation that is used in the mainstream nonlinear control literature [5, 56]. The method of input–output feedback linearisation for nonlinear MIMO control affine systems, which is central in this book, is presented in Section 3.1.3. Other linearisation techniques, such as exact linearisation and Volterra linearisation, are reviewed in Section 3.2. Finally, two feedback linearisation techniques that are applicable to systems that are not control affine are briefly presented in Section 3.3.

3.1 Nonlinear Control Affine Systems

Many nonlinear control methods are based on state space models where the time derivative of the states depends nonlinearly on the states and linearly on the control inputs [59]. Consider a multivariable nonlinear dynamic system with m inputs $\{u_1, \ldots, u_m\}$ and p outputs $\{y_1, \ldots, y_p\}$ described in a state space form by the following equations:

$$\dot{x} = f(x) + \sum_{j=1}^{m} g_j(x) u_j$$
$$y_i = h_i(x), i = 1, \ldots, p \qquad (3.1)$$

where $x = [x_1, \ldots, x_n]^T \in \Re^n$ is the state vector, $f(x), g_1(x), \ldots, g_m(x)$ are differentiable vector fields, and $h_1(x), \ldots, h_p(x)$ are smooth functions, all defined on an open set of $\Re^n$. This class of system is known as *control affine*, because the control input u appears linearly in the state equation. Equation 3.1 is used throughout this book in its compact form:

28   3. Introduction to Feedback Linearisation

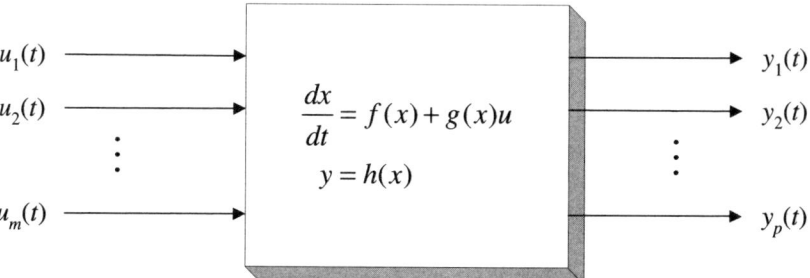

Fig. 3.1. Input affine nonlinear system

$$\dot{x} = f(x) + g(x)u$$
$$y = h(x) \tag{3.2}$$

with

$$f(x) = \begin{bmatrix} f_1(x_1, \ldots, x_n) \\ \vdots \\ f_n(x_1, \ldots, x_n) \end{bmatrix}_{n \times 1} \tag{3.3}$$

$$g(x) = \begin{bmatrix} g_1(x_1, \ldots, x_n), \cdots, g_m(x_1, \ldots, x_n) \end{bmatrix}_{n \times m} \tag{3.4}$$

$$h(x) = \begin{bmatrix} h_1(x_1, \ldots, x_n) \\ \vdots \\ h_p(x_1, \ldots, x_n) \end{bmatrix}_{p \times 1} \tag{3.5}$$

where $u = [u_1, \ldots, u_m]^T \in \Re^m$ is a vector of manipulated inputs and $y = [y_1, \ldots, y_p]^T \in \Re^p$ is a vector of output variables. A system of this kind is illustrated in Figure 3.1. In order to simplify the synthesis of the input–output linearising laws, the analysis in this book is restricted to systems having the same number of outputs and inputs ($p = m$), which are known as square systems.

Examples 3.1.1 and 3.1.2 illustrate the formulation of a class of control affine nonlinear systems of particular interest in this book, which are known as *dynamic neural networks*.

Example 3.1.1. Consider a system described by the following equations:

$$\dot{x}_1 = -\beta_1 x_1 + \omega_{11}\sigma(x_1) + \omega_{12}\sigma(x_2) + \gamma_1 u$$

$$\dot{x}_2 = -\beta_2 x_2 + \omega_{21}\sigma(x_1) + \omega_{22}\sigma(x_2) + \gamma_2 u \tag{3.6}$$

$$y = x_1$$

where $\sigma(\cdot) = \tanh(\cdot)$. This is an example of a SISO control affine system with $f(x)$, $g(x)$ and $h(x)$ given by:

3.1 Nonlinear Control Affine Systems

$$f(x) = \begin{bmatrix} -\beta_1 x_1 + w_{11}\sigma(x_1) + w_{12}\sigma(x_2) \\ -\beta_2 x_2 + w_{21}\sigma(x_1) + w_{22}\sigma(x_2) \end{bmatrix}$$

$$= - \underbrace{\begin{bmatrix} \beta_1 & 0 \\ 0 & \beta_2 \end{bmatrix}}_{\beta} \underbrace{\begin{bmatrix} x_1 \\ x_2 \end{bmatrix}}_{x} + \underbrace{\begin{bmatrix} w_{11} & w_{12} \\ w_{21} & w_{22} \end{bmatrix}}_{\omega} \underbrace{\begin{bmatrix} \sigma(x_1) \\ \sigma(x_2) \end{bmatrix}}_{\sigma(x)}$$

$$= -\beta x + \omega\sigma(x)$$

(3.7)

$$g(x) = \begin{bmatrix} \gamma_1 \\ \gamma_2 \end{bmatrix} = \gamma$$

$$h(x) = x_1$$

Notice that the vector fields $f(x)$, $g(x)$ and the function $h(x)$ are smooth and differentiable for all $x \in \Re^2$. ∎

Example 3.1.2. Consider the two-input two-output system described by the following equations:

$$\dot{x}_1 = -\beta_1 x_1 + w_{11}\sigma(x_1) + w_{12}\sigma(x_2) + w_{13}\sigma(x_3) + w_{14}\sigma(x_4) + \gamma_{11} u_1 + \gamma_{12} u_2$$

$$\dot{x}_2 = -\beta_2 x_2 + w_{21}\sigma(x_1) + w_{22}\sigma(x_2) + w_{23}\sigma(x_3) + w_{24}\sigma(x_4) + \gamma_{21} u_1 + \gamma_{22} u_2$$

$$\dot{x}_3 = -\beta_3 x_3 + w_{31}\sigma(x_1) + w_{32}\sigma(x_2) + w_{33}\sigma(x_3) + w_{34}\sigma(x_4) + \gamma_{31} u_1 + \gamma_{32} u_2$$

$$\dot{x}_4 = -\beta_4 x_4 + w_{41}\sigma(x_1) + w_{42}\sigma(x_2) + w_{43}\sigma(x_3) + w_{44}\sigma(x_4) + \gamma_{41} u_1 + \gamma_{42} u_2$$

$$y = [x_1 \quad x_2]^T$$

(3.8)

where $\sigma(\cdot) = \tanh(\cdot)$. This is an example of a MIMO control affine system with $f(x)$, $g(x)$ and $h(x)$ given by:

$$f(x) = \begin{bmatrix} -\beta_1 x_1 + w_{11}\sigma(x_1) + w_{12}\sigma(x_2) + w_{13}\sigma(x_3) + w_{14}\sigma(x_4) \\ -\beta_2 x_2 + w_{21}\sigma(x_1) + w_{22}\sigma(x_2) + w_{23}\sigma(x_3) + w_{24}\sigma(x_4) \\ -\beta_3 x_3 + w_{31}\sigma(x_1) + w_{32}\sigma(x_2) + w_{33}\sigma(x_3) + w_{34}\sigma(x_4) \\ -\beta_4 x_4 + w_{41}\sigma(x_1) + w_{42}\sigma(x_2) + w_{43}\sigma(x_3) + w_{44}\sigma(x_4) \end{bmatrix}$$

(3.9)

$$= - \underbrace{\begin{bmatrix} \beta_1 & 0 & 0 & 0 \\ 0 & \beta_2 & 0 & 0 \\ 0 & 0 & \beta_3 & 0 \\ 0 & 0 & 0 & \beta_4 \end{bmatrix}}_{\beta} \underbrace{\begin{bmatrix} x_1 \\ x_2 \\ x_3 \\ x_4 \end{bmatrix}}_{x} + \underbrace{\begin{bmatrix} w_{11} & w_{12} & w_{13} & w_{14} \\ w_{21} & w_{22} & w_{23} & w_{24} \\ w_{31} & w_{32} & w_{33} & w_{34} \\ w_{41} & w_{42} & w_{43} & w_{44} \end{bmatrix}}_{\omega} \underbrace{\begin{bmatrix} \sigma(x_1) \\ \sigma(x_2) \\ \sigma(x_3) \\ \sigma(x_4) \end{bmatrix}}_{\sigma(x)}$$

$$g(x) = \begin{bmatrix} \gamma_{11} & \gamma_{12} \\ \gamma_{21} & \gamma_{22} \\ \gamma_{31} & \gamma_{32} \\ \gamma_{41} & \gamma_{42} \end{bmatrix} = \gamma \qquad (3.10)$$

$$h(x) = [x_1 \quad x_2]^T$$

Notice that the vector fields $f(x)$, $g(x)$ and the function $h(x)$ are smooth and differentiable for all $x \in \Re^4$. ∎

3.1.1 Relative degree and characteristic matrix

The definitions of relative degree and characteristic matrix of multivariable systems are important for the development of feedback linearisation techniques [60, 61, 62].

Definition 3.1.1. *A multivariable nonlinear control affine system as defined in Equation 3.1, has a vector relative degree given by $\{r_1, r_2, \ldots, r_p\}$ at a point x_0 if*

1. $L_{g_i} L_f^k h_i(x) = 0$, $i = 1, \ldots, p$, $k = 0, \ldots, r_i - 2$ *for all x in a neighbourhood of x_0.*
2. *The characteristic matrix $C(x)$, given by*

$$C(x) = \begin{bmatrix} L_{g_1} L_f^{r_1-1} h_1(x) & L_{g_2} L_f^{r_1-1} h_1(x) & \cdots & L_{g_p} L_f^{r_1-1} h_1(x) \\ L_{g_1} L_f^{r_2-1} h_2(x) & L_{g_2} L_f^{r_2-1} h_2(x) & \cdots & L_{g_p} L_f^{r_2-1} h_2(x) \\ \vdots & \vdots & \ddots & \vdots \\ L_{g_1} L_f^{r_p-1} h_p(x) & L_{g_2} L_f^{r_p-1} h_p(x) & \cdots & L_{g_p} L_f^{r_p-1} h_p(x) \end{bmatrix}_{p \times p} \qquad (3.11)$$

is nonsingular at x_0. The total relative degree is defined as $r = r_1 + r_2 + \cdots + r_p$.

Example 3.1.3. Consider the model given by Equation 3.6, with $f(x)$, $g(x)$ and $h(x)$ given by Equation 3.7. We have that:

$$L_f^0 h(x) = h(x) = x_1 \qquad (3.12)$$

so that,

$$C(x) = L_g L_f^0 h(x) = \frac{\partial [x_1]}{\partial x} g(x) = [1 \quad 0] \begin{bmatrix} \gamma_1 \\ \gamma_2 \end{bmatrix} = \gamma_1 \qquad (3.13)$$

Therefore, using Definition 3.1.1, it is possible to conclude that the relative degree of this system is $r = 1$, provided $\gamma_1 \neq 0$.

If, however, $\gamma_1 = 0$, we have to evaluate $L_g L_f h(x)$:

3.1 Nonlinear Control Affine Systems   31

$$L_g L_f h(x) = L_g \left[\frac{\partial h(x)}{\partial x} f(x) \right]$$

$$= L_g \left([1\ 0] \begin{bmatrix} -\beta_1 x_1 + \omega_{11}\sigma(x_1) + \omega_{12}\sigma(x_2) \\ -\beta_2 x_2 + \omega_{21}\sigma(x_1) + \omega_{22}\sigma(x_2) \end{bmatrix} \right)$$

$$= L_g(-\beta_1 x_1 + \omega_{11}\sigma(x_1) + \omega_{12}\sigma(x_2))$$

$$= \frac{\partial(-\beta_1 x_1 + \omega_{11}\sigma(x_1) + \omega_{12}\sigma(x_2))}{\partial x} \begin{bmatrix} \gamma_1 \\ \gamma_2 \end{bmatrix}$$

$$= \underbrace{\gamma_1}_{=0} (-\beta_1 + \omega_{11}\sigma'(x_1)) + \gamma_2 \omega_{12}\sigma'(x_2)$$

$$= \gamma_2 \omega_{12} \sigma'(x_2)$$

(3.14)

where $\sigma'(x_2) = d\sigma(x_2)/dx_2$. In the case when $\gamma_1 = 0$ we can conclude that the relative degree of the system is $r = 2$, provided that $\gamma_2 \neq 0$ and $\omega_{12} \neq 0$. Notice that the derivative of $\tanh(z)$ becomes zero only when $z \to \pm\infty$. ∎

Example 3.1.4. Consider model given by Equation 3.8, with $f(x)$, $g(x)$ and $h(x)$ given by Equation 3.9. We have that:

$$L_f^0 h_1(x) = h_1(x) = x_1$$
$$L_f^0 h_2(x) = h_2(x) = x_2$$

(3.15)

so that,

$$L_{g_1} L_f^0 h_1(x) = [1\ 0\ 0\ 0] \begin{bmatrix} \gamma_{11} \\ \gamma_{21} \\ \gamma_{31} \\ \gamma_{41} \end{bmatrix} = \gamma_{11}$$

$$L_{g_2} L_f^0 h_1(x) = [1\ 0\ 0\ 0] \begin{bmatrix} \gamma_{12} \\ \gamma_{22} \\ \gamma_{32} \\ \gamma_{42} \end{bmatrix} = \gamma_{12}$$

$$L_{g_1} L_f^0 h_2(x) = [0\ 1\ 0\ 0] \begin{bmatrix} \gamma_{11} \\ \gamma_{21} \\ \gamma_{31} \\ \gamma_{41} \end{bmatrix} = \gamma_{21}$$

$$L_{g_2} L_f^0 h_2(x) = [0\ 1\ 0\ 0] \begin{bmatrix} \gamma_{12} \\ \gamma_{22} \\ \gamma_{32} \\ \gamma_{42} \end{bmatrix} = \gamma_{22}$$

(3.16)

If (i) $\gamma_{11} \neq 0$, (ii) $\gamma_{22} \neq 0$, and (iii) $\gamma_{11}\gamma_{22} - \gamma_{12}\gamma_{21} \neq 0$, then using Definition 3.1.1, it is possible to conclude that the vector relative degree of this system is $\{1, 1\}$, since in this case the characteristic matrix:

$$C(x) = \begin{bmatrix} L_{g_1} L_f^0 h_1(x) & L_{g_2} L_f^0 h_1(x) \\ L_{g_1} L_f^0 h_2(x) & L_{g_2} L_f^0 h_2(x) \end{bmatrix} = \begin{bmatrix} \gamma_{11} & \gamma_{12} \\ \gamma_{21} & \gamma_{22} \end{bmatrix} \qquad (3.17)$$

is nonsingular. The total relative degree of this system is $r = 1 + 1 = 2$. ∎

For SISO linear systems, the relative degree is the excess of poles over zeros. In a similar way, for nonlinear systems the relative degree r_i is the number of times the output $y_i(t)$ has to be differentiated in order to have at least one component of the input vector u explicitly appearing. In the same way, r_i is the minimum number of integrations that one component of the input vector has to go through to reach the $y_i(t)$ output. Each component of the vector relative degree is associated with its corresponding output. The relative degree concept is a decisive factor when defining the linearising feedback, because it defines the geometric order of the control signal for each output. The following formula describing the time derivatives of the output is an immediate consequences of Definition 3.1.1:

$$\frac{d^k y_i}{dt^k} = y_i^{(k)} = \begin{cases} L_f^k h_i(x), & k = 1, \ldots, r_i - 1 \\ L_f^k h_i(x) + \sum_{j=1}^{p} L_{g_j} L_f^{k-1} h_i(x) u_j, & k = r_i \end{cases} \qquad (3.18)$$

If a system output y_i and all its time derivatives do not depend explicitly on the external input u, the relative degree r_i for that output is not defined. This case of output-uncontrollable system is however considered atypical in well-formulated control problems. An interesting feature of the notion of relative degree is stated in the following proposition [5]. The concept of invariance of r_i under different conditions is a key element in the analysis and control structure of dynamic neural networks.

Proposition 3.1.1. *The vector relative degree $\{r_1, \ldots, r_p\}$ of a system of the form Equation 3.1 is invariant under coordinate transformations $z = [\xi^T, \eta^T]^T = \Phi(x)$ and under regular feedback $u = P(x) + Q(x)v$.*

Proof. Let $z = \Phi(x)$ be the coordinate transformation such that,

$$\bar{f}(z) = \left[\frac{\partial \Phi}{\partial x} f(x)\right]_{x=\Phi^{-1}(z)} \qquad (3.19a)$$

$$\bar{g}(z) = \left[\frac{\partial \Phi}{\partial x} g(x)\right]_{x=\Phi^{-1}(z)} \qquad (3.19b)$$

$$\bar{h}_i(z) = h_i(\Phi^{-1}(z)) \qquad (3.19c)$$

then

$$L_{\bar f}\bar h_i(z) = \frac{\partial \bar h_i}{\partial z}\bar f(z) = \left[\frac{\partial h_i}{\partial x}\right]_{x=\Phi^{-1}(z)}\left[\frac{\partial \Phi^{-1}}{\partial z}\right]\left[\frac{\partial \Phi}{\partial x}f(x)\right]_{x=\Phi^{-1}(z)} \quad (3.20a)$$

$$= \left[\frac{\partial h_i}{\partial x}f(x)\right]_{x=\Phi^{-1}(z)} \quad (3.20b)$$

$$= [L_f h_i(x)]_{x=\Phi^{-1}(z)} \quad (3.20c)$$

Iterated calculations of this kind show that

$$L_{\bar g}L_{\bar f}^k \bar h_i(z) = \left[L_g L_f^k h_i(x)\right]_{x=\Phi^{-1}(z)} \quad (3.21)$$

which shows that the relative degree is invariant under coordinate transformations. Concerning the invariance for a state feedback, consider a state feedback $u = P(x) + Q(x)v$, such that

$$\dot x = f(x) + g(x)u \quad (3.22)$$

becomes

$$\dot x = f(x) + g(x)[P(x) + Q(x)v] = [f(x) + g(x)P(x)] + [g(x)Q(x)]v \quad (3.23)$$

From Definition 3.1.1 it can be inductively proven that:

$$L_{f+gP}^k h_i(x) = L_f^k h_i(x), \quad i = 1,\ldots,p, \quad k = 0,\ldots,r_i - 1 \quad (3.24)$$

for all x near x_0. Moreover,

$$\left[L_{(gQ)_1} L_{f+gP}^{r_i-1} h_i(x^0) \ldots L_{(gQ)_p} L_{f+gP}^{r_i-1} h_i(x^0)\right] \quad (3.25)$$

$$= \left[L_{g_1} L_f^{r_i-1} h_i(x^0) \ldots L_{g_p} L_f^{r_i-1} h_i(x^0)\right] Q(x^0) \quad (3.26)$$

and thus, if the matrix $Q(x_0)$ is nonsingular:

$$\left[L_{(gQ)_1} L_{f+gP}^{r_i-1} h_i(x^0) \ldots L_{(gQ)_p} L_{f+gP}^{r_i-1} h_i(x^0)\right] \neq [0\ 0\ 0] \quad (3.27)$$

which proves the invariance of the vector relative degree under regular feedback. ∎

3.1.2 Zero dynamics and normal form

The notion of zero dynamics of a system refers to the internal dynamics of the states when the output $y(t)$ is forced to zero. Stability conditions of the zero dynamics are of considerable importance in the design of linearising control structures. For linear systems, the stability of the internal dynamics is simply determined by the location of the zeros in the transfer function. Analogously, an analysis of internal dynamics for nonlinear systems requires the extension of the concept of zeros. This task is not straightforward. First, a general transfer function representation for nonlinear systems does not exist. Second, specific stability conditions for nonlinear systems depend on the control input and the operating point. Keeping the output of the system identically zero uniquely represents an intrinsic property of a nonlinear system through the zero dynamics.

A good first approach is to transform the system described by Equation 3.2 into coordinates that facilitate the analysis of internal dynamics. Consider the transformation stated in the following proposition [5].

Proposition 3.1.2. *Suppose a system of the form of Equation 3.2 has a vector relative degree $\{r_1, \ldots, r_p\}$ at an operating point x_0. It is known that the individual relative degrees for the outputs are such that $r_1 + \cdots + r_p \leq n$. Define, for $1 \leq i \leq p$:*

$$\phi_1^i(x) = h_i(x)$$
$$\vdots \qquad (3.28)$$
$$\phi_{r_i}^i(x) = L_f^{r_i-1} h_i(x)$$

If $r = r_1 + \cdots + r_p$ is strictly less than n, it is always possible to find $n - r$ more functions $\varphi_{r+1}(x), \ldots, \varphi_n(x)$ such that the mapping

$$\Phi(x) = \left[\phi_1^1(x), \ldots, \phi_{r_1}^1(x), \ldots, \phi_1^p(x), \ldots, \phi_{r_p}^p(x), \phi_{r+1}(x), \ldots, \phi_n(x) \right]^T$$
$$(3.29)$$

represents a nonlinear transformation with a nonsingular Jacobian matrix around x_0. Moreover, if the distribution G generated by all possible combinations of $g_1, \ldots, g_p$ is involutive around x_0, the functions $\varphi_{r+1}(x), \ldots, \varphi_n(x)$ may be chosen to guarantee that

$$L_{g_j} \phi_i(x) = 0, \quad r+1 \leq i \leq n, \quad 1 \leq j \leq p \qquad (3.30)$$

The proof of this proposition can be found in [5]. Frobenius' Theorem 2.1.1 plays an essential role in its proof. Evidently, the case of $r_1 + \ldots + r_p > n$ is not feasible. If it takes more than n differentiations for the input to appear explicitly, the system is of order higher than n; if the control input never appears, the system is not controllable.

Differentiating Equation 3.28 with respect to time and using Equation 3.18, we obtain, for $i = 1, \ldots, p$:

$$\frac{d\phi_1^i(x)}{dt} = \frac{dh_i(x)}{dt} = L_f h_i(x) = \phi_2^i(x)$$
$$\vdots$$
$$\frac{d\phi_{r_i-1}^i(x)}{dt} = L_f^{r_i-1} h_i(x) = \phi_{r_i}^i(x) \qquad (3.31)$$
$$\frac{d\phi_{r_i}^i(x)}{dt} = L_f^{r_i} h_i(x) + \sum_{j=1}^p L_{g_j} L_f^{r_i-1} h_i(x) u_j$$

Separating the set of coordinates in,

$$\xi^i = \begin{bmatrix} \xi_1^i \\ \vdots \\ \xi_{r_i}^i \end{bmatrix} = \begin{bmatrix} \phi_1^i(x) \\ \vdots \\ \phi_{r_i}^i(x) \end{bmatrix}, \quad i = 1, \ldots, p \tag{3.32}$$

$$\xi = \begin{bmatrix} \xi^1 \\ \vdots \\ \xi^p \end{bmatrix} \tag{3.33}$$

$$\eta = \begin{bmatrix} \eta_1 \\ \vdots \\ \eta_{n-r} \end{bmatrix} = \begin{bmatrix} \phi_{r+1}(x) \\ \vdots \\ \phi_n(x) \end{bmatrix} \tag{3.34}$$

Then Equation 3.31 can be written as follows:

$$\begin{aligned} \frac{d\xi_1^i}{dt} &= \xi_2^i(t) \\ &\vdots \\ \frac{d\xi_{r_i-1}^i}{dt} &= \xi_{r_i}^i(t) \\ \frac{d\xi_{r_i}^i}{dt} &= b_i(\xi, \eta) + \sum_{j=1}^p a_{ij}(\xi, \eta) u_j \end{aligned} \tag{3.35}$$

where $y_i = \xi_1^i$, $i = 1, \ldots, p$ and

$$\begin{aligned} a_{ij}(\xi, \eta) &= L_{g_j} L_f^{r_i-1} h_i(\Phi^{-1}(\xi, \eta)) \\ b_i(\xi, \eta) &= L_f^{r_i} h_i(\Phi^{-1}(\xi, \eta)) \end{aligned} \tag{3.36}$$

Notice that $a_{ij}(x)$ is equal to the ij-element of the characteristic matrix $C(x)$ given by Equation 3.11. With respect to the last $n - r$ transformed variables, which correspond to $\eta = [\eta_1, \ldots, \eta_{n-r}]^T$, one can write the following expression for the general case where the distribution spanned by $g_1(x), \ldots, g_p(x)$ is not involutive:

$$\dot{\eta} = q(\xi, \eta) + \sum_{i=1}^p p_i(\xi, \eta) u_i = q(\xi, \eta) + p(\xi, \eta) u \tag{3.37}$$

Equations 3.35 and 3.37 constitute the *normal form* of a nonlinear affine system of the form given in Equation 3.1 with p inputs and p outputs and with a vector relative degree $\{r_1, \ldots, r_p\}$ around a point x_0.

3. Introduction to Feedback Linearisation

Grouping together the derivatives of $\xi^i_{r_i}$, $i = 1, \ldots, p$:

$$\begin{bmatrix} \frac{d\xi^1_{r_1}}{dt} \\ \vdots \\ \frac{d\xi^i_{r_i}}{dt} \\ \vdots \\ \frac{d\xi^p_{r_p}}{dt} \end{bmatrix} = \begin{bmatrix} y_1^{(r_1)} \\ \vdots \\ y_i^{(r_i)} \\ \vdots \\ y_p^{(r_p)} \end{bmatrix} = b(\xi(t), \eta(t)) + C(\xi(t), \eta(t))u(t) \qquad (3.38)$$

where $C(\xi(t), \eta(t))$ is the characteristic matrix 3.11 in (ξ, η) coordinates, that is, the original argument x has been replaced by $\Phi^{-1}(\xi, \eta)$, and $b(\xi, \eta)$ is given by:

$$b(\xi, \eta) = \begin{bmatrix} b_1(\xi, \eta) \\ \vdots \\ b_i(\xi, \eta) \\ \vdots \\ b_p(\xi, \eta) \end{bmatrix} = \begin{bmatrix} L_f^{r_1} h_1(\Phi^{-1}(\xi, \eta)) \\ \vdots \\ L_f^{r_i} h_i(\Phi^{-1}(\xi, \eta)) \\ \vdots \\ L_f^{r_p} h_p(\Phi^{-1}(\xi, \eta)) \end{bmatrix} \qquad (3.39)$$

If the outputs of the system are restricted to be zero, $y_i(t) = h_i(x) = \xi^i_1 = 0$, $i = 1, \ldots, p$, it follows that:

$$\xi^i_j = \Phi^i_j(x) = L_f^{j-1} h_i(x) = y_i^{(j-1)} = 0, \quad i = 1, \ldots, p, \quad j = 1, \ldots, r_i \qquad (3.40)$$

so that, from Equation 3.33, $\xi = 0$. If we also restrict the r_ith order time derivatives of the outputs to be zero, $y_i^{(r_i)} = 0$, $i = 1, \ldots, p$, then using Equation 3.38, it follows that the inputs are constrained to be the solution of the systems of equations:

$$0 = b(0, \eta) + C(0, \eta(t))u(t) \qquad (3.41)$$

Notice that by definition $C(x)$ is known to be nonsingular around x_0. If $f(x_0) = 0$, $h_i(x_0) = 0$, $i = 1, \ldots, p$, and $\phi_k(x_0) = 0$, $k = r + 1, \ldots, n$, then the normal form is defined around $(\xi, \eta) = (0, 0)$. In this case, it is reasonable to assume that $C(0, \eta(t))$ is nonsingular for small values of $\eta(t)$. Solving Equation 3.41 for u gives

$$u(t) = -[C(0, \eta(t))]^{-1} b(0, \eta(t)) \qquad (3.42)$$

Substituting Equation 3.42 into Equation 3.37 with $\xi = 0$, the expression so obtained characterises the *zero dynamics* of the system described by Equation 3.2:

$$\dot{\eta} = q_0(0, \eta) = q(0, \eta) - p(0, \eta)\left[C(0, \eta)\right]^{-1} b(0, \eta), \quad \eta(0) = \eta_0 \qquad (3.43)$$

Equation 3.43 characterises the internal dynamics of the system consistent with the restriction that the output vector is zero: $y(t) = 0$.

Remark 3.1.1 (Normal form for SISO systems). Consider a single-input single-output nonlinear system of the form

$$\dot{x} = f(x) + g(x)u$$
$$y = h(x) \qquad (3.44)$$

where $x \in \Re^n$, with relative degree $r < n$. Then its normal form around a point x^o, is given by:

$$\begin{aligned}\dot{\xi}_1 &= \xi_2 \\ \dot{\xi}_2 &= \xi_3 \\ &\vdots \\ \dot{\xi}_{r-1} &= \xi_r \\ \dot{\xi}_r &= b(\xi,\eta) + a(\xi,\eta)u \\ \dot{\eta} &= q(\xi,\eta) \\ y &= \xi_1\end{aligned} \qquad (3.45)$$

where

$$\xi = \begin{bmatrix} \xi_1 \\ \xi_2 \\ \vdots \\ \xi_r \end{bmatrix} \qquad (3.46)$$

$$\eta = \begin{bmatrix} \eta_1 \\ \eta_2 \\ \vdots \\ \eta_{n-r} \end{bmatrix} \qquad (3.47)$$

$$a(\xi,\eta) = L_g L_f^{r-1} h(\Phi^{-1}(\xi,\eta)) \qquad (3.48)$$
$$b(\xi,\eta) = L_f^r h(\Phi^{-1}(\xi,\eta)) \qquad (3.49)$$

Example 3.1.5. Consider the system described by Equation 3.6, with $f(x)$, $g(x)$ and $h(x)$ given by Equation 3.7 and assume that $\gamma_1 \neq 0$. In this case, as seen in Example 3.1.3, the relative degree of this system is $r = 1$. In order to obtain the normal form, set:

$$\xi = \phi_1(x) = h(x) = x_1 \qquad (3.50)$$
$$\qquad (3.51)$$

We know from Equation 3.30 that $\phi_2(x)$ satisfies the following partial differential equation:

$$L_g \phi_2(x) = 0 \Rightarrow \gamma_1 \frac{\partial \phi_2}{\partial x_1} + \gamma_2 \frac{\partial \phi_2}{\partial x_2} = 0 \qquad (3.52)$$

3. Introduction to Feedback Linearisation

If we take

$$\phi_2 = -\frac{-\gamma_1 x_2 + \gamma_2 x_1}{\gamma_1} \tag{3.53}$$

it is easy to see that Equation 3.52 is satisfied. Therefore, the coordinate transformation is linear and given by:

$$\Phi(x) = \begin{bmatrix} x_1 \\ -\frac{-\gamma_1 x_2 + \gamma_2 x_1}{\gamma_1} \end{bmatrix} = \begin{bmatrix} 1 & 0 \\ -\frac{\gamma_2}{\gamma_1} & 1 \end{bmatrix} \begin{bmatrix} x_1 \\ x_2 \end{bmatrix} \tag{3.54}$$

The Jacobian of this transformation is:

$$\frac{\partial \Phi(x)}{\partial x} = \begin{bmatrix} 1 & 0 \\ -\frac{\gamma_2}{\gamma_1} & 1 \end{bmatrix} \tag{3.55}$$

which is constant and nonsingular. The inverse transformation is given by:

$$\begin{bmatrix} x_1 \\ x_2 \end{bmatrix} = \begin{bmatrix} 1 & 0 \\ \frac{\gamma_2}{\gamma_1} & 1 \end{bmatrix} \begin{bmatrix} \xi \\ \eta \end{bmatrix} = \begin{bmatrix} \xi \\ \frac{\gamma_1 \eta + \gamma_2 \xi}{\gamma_1} \end{bmatrix} \tag{3.56}$$

In the new coordinates (ξ, η), the system in Equation 3.6 can be expressed as follows:

$$\begin{aligned} \dot{\xi} &= -\beta_1 \xi + \omega_{11} \sigma(\xi) + \omega_{12} \sigma\left(\frac{\gamma_1 \eta + \gamma_2 \xi}{\gamma_1}\right) + \gamma_1 u \\ \dot{\eta} &= \frac{\gamma_2}{\gamma_1}(\beta_1 - \beta_2)\xi - \beta_2 \eta + \left(\omega_{21} - \frac{\gamma_2}{\gamma_1}\omega_{11}\right)\sigma(\xi) \\ &\quad + \left(\omega_{22} - \frac{\gamma_2}{\gamma_1}\omega_{12}\right)\sigma\left(\frac{\gamma_1 \eta + \gamma_2 \xi}{\gamma_1}\right) \end{aligned} \tag{3.57}$$

$$y = \xi$$

The input that makes the output zero is thus given by:

$$u(t) = -\frac{\omega_{12}\sigma(\eta(t))}{\gamma_1} \tag{3.58}$$

And finally, the zero dynamics are described by:

$$\dot{\eta} = -\beta_2 \eta + \left(\omega_{22} - \frac{\gamma_2}{\gamma_1}\omega_{12}\right)\sigma(\eta) \tag{3.59}$$

∎

3.1 Nonlinear Control Affine Systems

Stability of the zero dynamics. The stability of the zero dynamics and its asymptotic properties play an important role in feedback linearisation, as we will see in Section 3.1.5. Consider now the following example where the conditions for the stability of the zero dynamics are studied.

Example 3.1.6. Consider the zero dynamics equation obtained in Example 3.1.5:

$$\dot{\eta} = -\beta_2 \eta + \left(\omega_{22} - \frac{\gamma_2}{\gamma_1}\omega_{12}\right)\sigma(\eta) \qquad (3.60)$$

This is an autonomous system with (at least) an equilibrium point at $\eta = 0$. Consider the Lyapunov function candidate $V(\eta) = \eta^2$. Then we have:

$$\dot{V}(\eta) = \frac{\partial V}{\partial \eta}\dot{\eta} = 2\eta\left[-\beta_2\eta + \left(\omega_{22} - \frac{\gamma_2}{\gamma_1}\omega_{12}\right)\sigma(\eta)\right] \qquad (3.61)$$

$$= -2\beta_2\eta^2 + 2\left(\omega_{22} - \frac{\gamma_2}{\gamma_1}\omega_{12}\right)\eta\sigma(\eta) \qquad (3.62)$$

We infer that $\dot{V}(\eta)$ is negative definite provided that $\beta_2 > 0$ and $(\omega_{22} - \gamma_2/\gamma_1 \omega_{12}) \leq 0$. So, by Theorem 2.2.2, we can say that if the following conditions are satisfied:

1. $\beta_2 > 0$
2. $\omega_{22} \leq \frac{\gamma_2}{\gamma_1}\omega_{12}$

then $\eta = 0$ is an asymptotically stable equilibrium point of the zero dynamics found in Example 3.1.5. ∎.

3.1.3 Input–output linearisation

The objective of input–output linearisation is to introduce a new input variable v and a nonlinear transformation that uses state feedback to compute the original input u, such that a system described by Equation 3.2 behaves linearly from the new input v to the output y. This is achieved without requiring any conditions on the internal dynamics of the system apart from their stability.

Definition 3.1.2. *(Kravaris and Soroush, [63]) A MIMO nonlinear system with p inputs and p outputs of the form shown in Equation 3.2 is called input–output linearisable if there exists a static state feedback of the form,*

$$u = P(x) + Q(x)v \qquad (3.63)$$

with $P(x) \in \Re^p$, $Q(x) \in \Re^{p \times p}$ nonsingular, $v \in \Re^p$ an external input vector; and linear operators of the form:

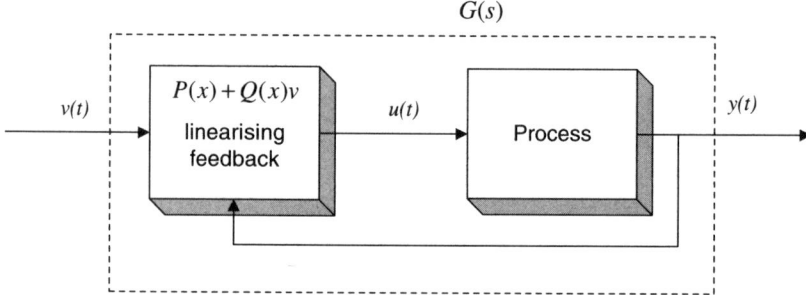

Fig. 3.2. Multi-input multi-output feedback linearising structure

$$\mathfrak{L}_{\rho_i} = \sum_{k=0}^{\rho_i} \lambda_{ik} \frac{d^k}{dt^k}, \quad i=1,\ldots,p \tag{3.64}$$

with constant coefficients $\lambda_{ik} = [\lambda_{ik}^1 \lambda_{ik}^2 \ldots \lambda_{ik}^p]^T \in \Re^p$ satisfying $\lambda_{i\rho_i} \neq 0$, ρ_i the order of the operator and

$$\det\left(\left[\left(\sum_{k=0}^{\rho_1}\lambda_{1k}s^k\right)\left(\sum_{k=0}^{\rho_2}\lambda_{2k}s^k\right)\ldots\left(\sum_{k=0}^{\rho_p}\lambda_{pk}s^k\right)\right]_{p\times p}\right) \neq 0 \tag{3.65}$$

such that

$$\sum_{i=1}^{p}\mathfrak{L}_{\rho_i}y_i = v \tag{3.66}$$

The Laplace domain matrix fraction description of the system after it has been linearised according to Definition 3.1.2 is given by:

$$y(s) = G(s)v(s) = [B(s)]^{-1} v(s) \tag{3.67}$$

where

$$B(s) = \left[\left(\sum_{k=0}^{\rho_1}\lambda_{1k}s^k\right)\left(\sum_{k=0}^{\rho_2}\lambda_{2k}s^k\right)\ldots\left(\sum_{k=0}^{\rho_p}\lambda_{pk}s^k\right)\right] \tag{3.68}$$

and $y(s)$ and $v(s)$ denote the Laplace transforms of $y(t)$ and $v(t)$, respectively. The feedback linearisation structure is illustrated in Figure 3.2.

At this point it is important to characterise the systems that can be input–output linearised. The following Proposition and Theorem provide necessary and sufficient conditions for the existence of a feedback law given by Equation 3.63 that is able to input–output linearise a control affine nonlinear system [63].

3.1 Nonlinear Control Affine Systems

Proposition 3.1.3. *(Kravaris and Soroush, [63]) A necessary condition for a system of the form given in Equation 3.2 to be input–output linearisable according to Definition 3.1.2 is that each output y_i has a relative degree associated to it, as given in Definition 3.1.1. Additionally, if ρ_i are the orders of the linear operators $\mathfrak{L}_{\rho_i}$ and the vector relative degree is $\{r_1, \ldots, r_p\}$, then $\rho_i \geqslant r_i$, $i = 1, \ldots, p$.*

Proof. Suppose that the output y_i has its relative degree not defined, i.e. all derivatives dy_i^k/dt^k are independent of u for all k. After the linearising feedback $u = P(x) + Q(x)v$ is applied, the derivatives dy_i^k/dt^k are independent of v for all k, and the ith row of the matrix:

$$\left[\left(\sum_{k=0}^{\rho_1} \lambda_{1k} s^k\right)\left(\sum_{k=0}^{\rho_2} \lambda_{2k} s^k\right) \cdots \left(\sum_{k=0}^{\rho_p} \lambda_{pk} s^k\right)\right]^{-1} \quad (3.69)$$

is a null vector. This contradicts Equation 3.65. Analogously, the assumption of $\rho_i < r_i$ leads to contradiction. □

Proposition 3.1.3 provides a necessary condition for input–output linearisability: all outputs must have relative degrees. The following Theorem provides a state feedback law that makes the orders of the linear operators minimal: $\rho_i = r_i, i = 1, \ldots, p$.

Theorem 3.1.1. *(Kravaris and Soroush, [63]) For arbitrary $\lambda_{ik} \in \Re^n$ ($k = 0, \ldots, r_i$; $i = 1, \ldots, p$), a state feedback of the form in Equation 3.63 defined by*

$$\begin{aligned} u(t) = &\left(\begin{bmatrix} \lambda_{1r_1} & \cdots & \lambda_{pr_p} \end{bmatrix}_{p \times p} C(x)\right)^{-1}_{p \times p} \\ &\times \left(v - \sum_{i=1}^{p} \sum_{k=0}^{r_i} \lambda_{ik} L_f^k h_i(x)\right)_{p \times 1} \end{aligned} \quad (3.70)$$

applied to the system described by Equation 3.2, where $C(x)$ is the characteristic matrix of the system given in Equation 3.11, produces a linearised input–output description given by:

$$\sum_{i=1}^{p} \sum_{k=0}^{r_i} \lambda_{ik} \frac{d^k y_i}{dt^k} = v \quad (3.71)$$

when the following sufficient conditions hold:

- *The vector relative degree is well defined.*
- *λ_{ik} satisfies*

42   3. Introduction to Feedback Linearisation

$$\det\left(\left[\left(\sum_{k=0}^{r_1}\lambda_{1k}s^k\right)\left(\sum_{k=0}^{r_2}\lambda_{2k}s^k\right)\cdots\left(\sum_{k=0}^{r_p}\lambda_{pk}s^k\right)\right]_{p\times p}\right)\neq 0 \quad (3.72\text{a})$$

$$\det\left(\begin{bmatrix}\lambda_{1r_1} & \lambda_{2r_2} & \cdots & \lambda_{pr_p}\end{bmatrix}\right)\neq 0 \quad (3.72\text{b})$$

Proof. Consider a system of the form given in Equation 3.2 with a vector relative degree $\{r_1,\ldots,r_p\}$. By using Equation 3.18 for the time derivatives of the output y, the following expression is obtained,

$$\begin{aligned}
\sum_{i=1}^{p}\sum_{k=0}^{r_i}\lambda_{ik}\frac{d^k y_i}{dt^k} &= \sum_{i=1}^{p}\left[\sum_{k=0}^{r_i-1}\lambda_{ik}\frac{d^k y_i}{dt^k} + \lambda_{ir_i}\frac{d^{r_i} y_i}{dt^{r_i}}\right]\\
&= \sum_{i=1}^{p}\sum_{k=0}^{r_i}\lambda_{ik}L_f^k h_i(x) + \sum_{i=1}^{p}\lambda_{ir_i}\left[\sum_{j=1}^{p}L_{g_j}L_f^{r_i-1}h_i(x)u_j\right]\\
&= \sum_{i=1}^{p}\sum_{k=0}^{r_i}\lambda_{ik}L_f^k h_i(x) + \sum_{i=1}^{p}\lambda_{ir_i}\begin{bmatrix}L_{g_1}L_f^{r_i-1}h_i(x) & \cdots & L_{g_p}L_f^{r_i-1}h_i(x)\end{bmatrix}\begin{bmatrix}u_1\\ \vdots \\ u_p\end{bmatrix}\\
&= \sum_{i=1}^{p}\sum_{k=0}^{r_i}\lambda_{ik}L_f^k h_i(x) + \begin{bmatrix}\lambda_{1r_1} & \cdots & \lambda_{pr_p}\end{bmatrix}\\
&\quad \times \begin{bmatrix}L_{g_1}L_f^{r_1-1}h_1(x) & \cdots & L_{g_p}L_f^{r_1-1}h_1(x)\\ \vdots & \ddots & \vdots\\ L_{g_1}L_f^{r_p-1}h_p(x) & \cdots & L_{g_p}L_f^{r_p-1}h_p(x)\end{bmatrix}\begin{bmatrix}u_1\\ \vdots \\ u_p\end{bmatrix}\\
&= \sum_{i=1}^{p}\sum_{k=0}^{r_i}\lambda_{ik}L_f^k h_i(x) + \begin{bmatrix}\lambda_{1r_1} & \cdots & \lambda_{pr_p}\end{bmatrix}C(x)u
\end{aligned} \quad (3.73)$$

where $C(x)$ is the characteristic matrix of the system (see Equation 3.11). The nonsingularity of the characteristic matrix together with Equation 3.72b guarantee the existence of the matrix:

$$\left[\begin{bmatrix}\lambda_{1r_1} & \cdots & \lambda_{pr_p}\end{bmatrix}_{p\times p}C(x)\right]_{p\times p}^{-1} \quad (3.74)$$

Therefore, substituting u by the expression in Equation 3.70, makes the right-hand side of Equation 3.73 equal to v. The state feedback given by Equation 3.70 makes the system input–output linear in the form of Equation 3.71 and in the sense of Definition 3.1.2. □

Remark 3.1.2. Notice that the linearising control law in Equation 3.70 can be written as follows:

$$u(t) = P(x) + Q(x)v(t) \quad (3.75)$$

where

3.1 Nonlinear Control Affine Systems

$$P(x) = -\left(\begin{bmatrix} \lambda_{1r_1} & \ldots & \lambda_{pr_p} \end{bmatrix}_{p\times p} C(x)\right)^{-1}_{p\times p} \left(\sum_{i=1}^{p}\sum_{k=0}^{r_i} \lambda_{ik} L_f^k h_i(x)\right)_{p\times 1} \quad (3.76)$$

and

$$Q(x) = \left(\begin{bmatrix} \lambda_{1r_1} & \ldots & \lambda_{pr_p} \end{bmatrix}_{p\times p} C(x)\right)^{-1}_{p\times p} \quad (3.77)$$

Remark 3.1.3. In the case of a SISO system with relative degree r, $C(x) = L_g L_f^{r-1} h(x) \neq 0$, the linearising input is given by:

$$u = \frac{v - \sum_{k=0}^{r} \lambda_k L_f^k h(x)}{\lambda_r L_g L_f^{r-1} h(x)} \quad (3.78)$$

and the linearised description of the system is given by:

$$\sum_{k=0}^{r} \lambda_k \frac{d^k y}{dt^k} = v \quad (3.79)$$

which has the equivalent Laplace domain representation:

$$y(s) = \frac{1}{\lambda_r s^r + \cdots + \lambda_0} v(s) \quad (3.80)$$

Notice that the Laplace domain representation of the input–output linearised system given by Equation 3.80 has no finite zeros and its poles are given by the design parameters $\lambda_0, \ldots, \lambda_\rho$. Also notice that in (ξ, η) coordinates, the linearising control law in Equation 3.78 is given by:

$$u = \frac{1}{a(\xi, \eta)}(-b(\xi, \eta) - \lambda_0 \xi_1 - \lambda_1 \xi_2 - \cdots - \lambda_{r-1} \xi_r) \quad (3.81)$$

with $a(\xi, \eta)$ and $b(\xi, \eta)$ given by Equations 3.48 and 3.49, respectively. Therefore, the normal form of the input–output linearised system is given by:

$$\begin{aligned}
\dot{\xi}_1 &= \xi_2 \\
\dot{\xi}_2 &= \xi_3 \\
&\vdots \\
\dot{\xi}_{r-1} &= \xi_r \\
\dot{\xi}_r &= -\lambda_0 \xi_1 - \lambda_1 \xi_2 - \cdots - \lambda_{r-1} \xi_r + v \\
\dot{\eta} &= q(\xi, \eta) \\
y &= \xi_1
\end{aligned} \quad (3.82)$$

44   3. Introduction to Feedback Linearisation

which can be written in a shorter form as follows:

$$\dot{\xi} = A\xi + Bv$$

$$\dot{\eta} = q(\xi, \eta) \tag{3.83}$$

$$y = \xi_1$$

where

$$A = \begin{bmatrix} 0 & 1 & 0 & \cdots & 0 \\ 0 & 0 & 1 & \cdots & 0 \\ \vdots & \vdots & \vdots & \ddots & \vdots \\ 0 & 0 & 0 & \cdots & 1 \\ -\lambda_0 & -\lambda_1 & -\lambda_2 & \cdots & -\lambda_{r-1} \end{bmatrix} \tag{3.84}$$

$$B = \begin{bmatrix} 0 \\ 0 \\ \vdots \\ 1 \end{bmatrix} \tag{3.85}$$

Remark 3.1.4. If the characteristic matrix $C(x)$ is not singular but close to singular (for example, if it has a small determinant or a large condition number [64]), it may be a good idea to design the linearising law using $\rho_i > r_i$ instead of r_i, at least on some of the outputs (even though this would result in an approximate input–output linearisation). ρ_i, $i = 1, \ldots, p$, may then be seen as design parameters. One of the paramount issues to check is that the linearising control signal $u(t)$ does not attain values outside its permissible limits. If $C(x)$ is close to singular, then the values of the linearising control signal $u(t)$ can be very large when $r_i, i = 1, \ldots p$, is used to design the linearising law.

Example 3.1.7. Consider the SISO system of Example 3.1.1:

$$\dot{x}_1 = -\beta_1 x_1 + \omega_{11}\sigma(x_1) + \omega_{12}\sigma(x_2) + \gamma_1 u$$

$$\dot{x}_2 = -\beta_2 x_2 + \omega_{21}\sigma(x_1) + \omega_{22}\sigma(x_2) + \gamma_2 u \tag{3.86}$$

$$y = x_1$$

where $\sigma(\cdot) = \tanh(\cdot)$, and suppose that $\gamma_1 \neq 0$, so that its relative degree is $r = 1$ (see Example 3.1.3). Suppose that λ_0 and λ_1 are the design parameters for the linearised system. Then, from Equation 3.78, the linearising input is given by:

3.1 Nonlinear Control Affine Systems   45

$$u = \frac{v - \sum_{k=0}^{r} \lambda_k L_f^k h(x)}{\lambda_r L_g L_f^{r-1} h(x)} = \frac{v - \lambda_0 L_f^0[x_1] - \lambda_1 L_f^1[x_1]}{\lambda_1 L_g L_f^0[x_1]} \quad (3.87)$$

$$= \frac{v - \lambda_0 x_1 - \lambda_1(-\beta_1 x_1 + \omega_{11}\sigma(x_1) + \omega_{12}\sigma(x_2))}{\lambda_1 \gamma_1}$$

Applying the linearising law in Equation 3.87 to the first differential equation of 3.86, gives:

$$\dot{x}_1 = \frac{v - \lambda_0 x_1}{\lambda_1} \quad (3.88)$$

Replacing x_1 by y above and re-arranging the terms, gives:

$$\lambda_1 \dot{y} + \lambda_0 y = v \quad (3.89)$$

So that the input–output behaviour of the system from the external input v to the output y is given in the Laplace domain by:

$$y(s) = \frac{1}{\lambda_1 s + \lambda_0} v(s) \quad (3.90)$$

Notice that the linearised input–output dynamics are of first order. Suppose now that $\gamma_1 = 0$, $\gamma_2 \neq 0$ and $\omega_{12} \neq 0$, so that the relative degree of the system is $r = 2$. Suppose that the design parameters are λ_0, λ_1 and λ_2. Then, from Equation 3.78, the linearising input is given by:

$$u = \frac{v - \sum_{k=0}^{r} \lambda_k L_f^k h(x)}{\lambda_r L_g L_f^{r-1} h(x)} = \frac{v - \lambda_0 L_f^0[x_1] - \lambda_1 L_f^1[x_1] - \lambda_2 L_f^2[x_1]}{\lambda_2 L_g L_f^1[x_1]}$$

$$= \frac{v - \lambda_0 x_1 - \lambda_1(-\beta_1 x_1 + \omega_{11}\sigma(x_1) + \omega_{12}\sigma(x_2))}{D} \quad (3.91)$$

$$- \frac{\lambda_2(-\beta_1 x_1 + \omega_{11}\sigma(x_1) + \omega_{12}\sigma(x_2))(-\beta_1 + \omega_{11}\sigma'(x_1))}{D}$$

$$- \frac{\lambda_2(-\beta_2 x_2 + \omega_{21}\sigma(x_1) + \omega_{22}\sigma(x_2))\omega_{12}\sigma'(x_2)}{D}$$

where

$$D = \lambda_2 \big(\underbrace{\gamma_1}_{=0}(-\beta_1 + \omega_{11}\sigma'(x_1)) + \gamma_2 \omega_{12}\sigma'(x_2)\big) \quad (3.92)$$

$$= \lambda_2 \gamma_2 \omega_{12}\sigma'(x_2)$$

Applying the linearising law in Equation 3.91 to the system given by 3.86, gives the following description:

$$\lambda_2 \ddot{y} + \lambda_1 \dot{y} + \lambda_0 y = v \quad (3.93)$$

So that the input–output behaviour of the system from the external input v to the output y is given in the Laplace domain by:

$$y(s) = \frac{1}{\lambda_2 s^2 + \lambda_1 s + \lambda_0} v(s) \quad (3.94)$$

■

3.1.4 Input–output linearisation and decoupling

An approach used in this book is input–output linearisation and decoupling, which is a specific case of input–output linearisation. An appropriate selection of the design parameters λ_{ij} in Equation 3.70 leads to a feedback linearised system where the ith output depends on the ith external input only, having a particular case of Equation 3.71. Let the following Theorem present a description of the linearising-decoupling technique [63].

Theorem 3.1.2. *For arbitrary values $\hat{\lambda}_{ik}$ ($i = 1, \ldots, p$ and $k = 0, \ldots, r_i$) a state feedback*

$$u = P(x) + Q(x)v \qquad (3.95)$$

where

$$P(x) = -A(x)^{-1}B(x) \qquad (3.96a)$$
$$Q(x) = A(x)^{-1} \qquad (3.96b)$$

$$A(x) = \begin{bmatrix} \hat{\lambda}_{1r_1} & 0 & \cdots & 0 \\ 0 & \hat{\lambda}_{2r_2} & \cdots & 0 \\ \vdots & \vdots & \ddots & \vdots \\ 0 & 0 & \cdots & \hat{\lambda}_{pr_p} \end{bmatrix} C(x) \qquad (3.96c)$$

$$= \begin{bmatrix} \hat{\lambda}_{1r_1} L_{g_1} L_f^{r_1-1} h_1(x) & \cdots & \hat{\lambda}_{1r_1} L_{g_p} L_f^{r_1-1} h_1(x) \\ \vdots & \ddots & \vdots \\ \hat{\lambda}_{pr_p} L_{g_1} L_f^{r_p-1} h_p(x) & \cdots & \hat{\lambda}_{pr_p} L_{g_p} L_f^{r_p-1} h_p(x) \end{bmatrix}_{p \times p} \qquad (3.96d)$$

$$B(x) = \begin{bmatrix} \sum_{k=0}^{r_1} \hat{\lambda}_{1k} L_f^k h_1(x) \\ \vdots \\ \sum_{k=0}^{r_p} \hat{\lambda}_{pk} L_f^k h_p(x) \end{bmatrix}_{p \times 1} \qquad (3.96e)$$

where $\hat{\lambda}_{ik}$ are scalar design parameters, $C(x)$ is the characteristic matrix given in Equation 3.11 and r_i is the relative degree of the ith output y_i, produces, when applied to the system in Equation 3.2, a linearised and decoupled input–output description given by,

$$\sum_{k=0}^{r_i} \hat{\lambda}_{ik} \frac{d^k y_i}{dt^k} = v_i, \quad i = 1 \ldots p \qquad (3.97)$$

when the conditions from Theorem 3.1.1 are fulfilled.

Proof. Taking the vectors $e_1, \ldots, e_p$,

3.1 Nonlinear Control Affine Systems

$$e_1 = \begin{bmatrix} 1 & 0 & \cdots & 0 \end{bmatrix}^T_{1\times p}, \; e_2 = \begin{bmatrix} 0 & 1 & \cdots & 0 \end{bmatrix}^T_{1\times p}, \; \ldots, \; e_p = \begin{bmatrix} 0 & 0 & \cdots & 1 \end{bmatrix}^T_{1\times p} \quad (3.98)$$

and defining λ_{ik} such that

$$\lambda_{ik} = \hat{\lambda}_{ik} e_i, \quad i = 1 \ldots p, \; k = 0 \ldots r_i \quad (3.99)$$

Given that $\hat{\lambda}_{ir_i} \neq 0$, $i = 1, \ldots, p$, from Theorem 3.1.1, then there exist design parameters $\hat{\lambda}_{ik}$ and thus λ_{ik} vectors for which the matrix in Equation 3.74 is nonsingular.

Now, taking the time derivatives from Equation 3.18, the left side of Equation 3.97 is given by,

$$\sum_{k=0}^{r_i} \lambda_{ik} \frac{d^k y_i}{dt^k} = \sum_{k=0}^{r_i-1} \left[\hat{\lambda}_{ik} L_f^k h_i(x) \right] + \hat{\lambda}_{ir_i} L_f^{r_i} h_i(x) + \hat{\lambda}_{ir_i} \sum_{j=1}^{p} L_{g_j} L_f^{r_i-1} h_i(x) u_j \quad (3.100)$$

and from Equation 3.95, the new external input v is given by,

$$v = A(x)u + B(x)$$

$$= \begin{bmatrix} \hat{\lambda}_{1r_1} L_{g_1} L_f^{r_1-1} h_1(x) u_1 + \ldots + \hat{\lambda}_{1r_1} L_{g_p} L_f^{r_1-1} h_1(x) u_p + \sum_{k=0}^{r_1} \hat{\lambda}_{1k} L_f^k h_1(x) \\ \vdots \\ \hat{\lambda}_{pr_p} L_{g_1} L_f^{r_p-1} h_p(x) u_1 + \ldots + \hat{\lambda}_{pr_p} L_{g_p} L_f^{r_p-1} h_p(x) u_p + \sum_{k=0}^{r_p} \hat{\lambda}_{pk} L_f^k h_p(x) \end{bmatrix} \quad (3.101)$$

where the ith component is such that,

$$v_i = \hat{\lambda}_{ir_i} L_{g_1} L_f^{r_i-1} h_i(x) u_1 + \ldots + \hat{\lambda}_{ir_i} L_{g_p} L_f^{r_i-1} h_i(x) u_p + \sum_{k=0}^{r_i} \hat{\lambda}_{ik} L_f^k h_i(x)$$

$$= \hat{\lambda}_{ir_i} \sum_{j=1}^{p} L_{g_j} L_f^{r_i-1} h_i(x) u_j + \sum_{k=0}^{r_i} \hat{\lambda}_{ik} L_f^k h_i(x) \quad (3.102)$$

As the right-hand sides of Equations 3.100 and 3.102 are the same, then the system obeys Equation 3.97. It follows that the control signal given by Equation 3.95 input–output linearises and decouples the system described by Equation 3.2. □

Input–output decoupling is a specific case of input–output linearisation. The conditions for a nonlinear system to be decoupable by means of state feedback are the existence of a vector relative degree and the nonsingularity of the characteristic matrix. Hence, a system is decoupable whenever it is linearisable.

The linearising-decoupling case is easier to implement than the general linearising technique because of the simplicity of the vectors in Equation 3.98. The final linearising control law given by Equation 3.95 requires far fewer operations than the general form shown in Equation 3.70, substantially reducing the computational cost of the implementation. Furthermore, the reduction of the interactions of the controlled variables increases the viability of linear SISO controllers.

On the other hand, decoupled closed loop responses are not always the best choice. There are cases in which decoupling control signals force the intrinsic structural constraints of the plant, deteriorating its response. Additionally, it could demand large control signals in its attempt to weaken the interactions. When approximated models are used to represent the plant, the closed loop response of the plant is not perfectly decoupled. Effectively, the objective is to weaken the interaction between output variables and not disconnect them completely, which is possible but not necessarily appropriate [65].

Another important aspect of the linearising-decoupling control law is its validity over extensive operating regions. Despite the fact that the control law is described using a vector relative degree defined on an operating point, the achievement of an input–output noninteractive behaviour is independent of the coordinates used in the state space description.

3.1.5 Stability of input–output linearised systems

The synthesis of the input–output linearising controller involves the pole placement for the closed loop system. Therefore, the input–output stability of the linearised system depends on the roots of its characteristic equation. As a result, the design parameters of the linearising control signal can always be chosen for closed loop stability and fast dynamics. The following Definition and Proposition clearly state this condition.

Definition 3.1.3. *(see [66]) A nonsingular polynomial matrix $D(s)$ with p rows and p columns is said to be column reduced if*

$$\deg \det D(s) = \sum_{i=1}^{p} \delta_{ci} \tag{3.103}$$

where δ_{ci} is the degree of the ith column of $D(s)$, which is defined as the highest power of s in the ith column of $D(s)$.

Proposition 3.1.4. *The input–output linearised form of system 3.2, given by Equation 3.71, is bounded-input-bounded-output (BIBO) stable if the roots of the characteristic equation*

3.1 Nonlinear Control Affine Systems

$$\det\left(\left[\left(\sum_{k=0}^{r_1}\lambda_{1k}s^k\right)\left(\sum_{k=0}^{r_2}\lambda_{2k}s^k\right)\cdots\left(\sum_{k=0}^{r_p}\lambda_{pk}s^k\right)\right]_{p\times p}\right)=0 \quad (3.104)$$

have negative real part.

Proof. The matrix fraction description of the linearised closed loop system in the Laplace domain given by Equation 3.67 satisfies the condition 3.65. From condition 3.72b, it can be equivalently stated that the matrix

$$\left[\left(\sum_{k=0}^{r_1}\lambda_{1k}s^k\right)\left(\sum_{k=0}^{r_2}\lambda_{2k}s^k\right)\cdots\left(\sum_{k=0}^{r_p}\lambda_{pk}s^k\right)\right]_{p\times p} \quad (3.105)$$

is column-reduced with column degrees $r_1,\ldots,r_p$, respectively. The determinant of the reduced form of Equation 3.65 is the characteristic polynomial of the input–output linearised system given by Equation 3.104.

Given that the input–output dynamics are linear,

$$\sum_{i=1}^{p}\sum_{k=0}^{r_i}\lambda_{ik}\frac{d^k y_i}{dt^k}=v \quad (3.106)$$

then from linear control theory, a system is BIBO stable when the roots of the characteristic polynomial are on the left-hand side of the imaginary plane [66]. □

Apart from input–output stability, it is relevant to study conditions for the internal stability of the $v-y$ system. The stability of internal dynamics is given by the boundedness of the solution of Equation 3.2 under the linearising control law 3.70 with no external input ($v=0$).

When a system described by Equation 3.2 is linearised by the state feedback given by Equation 3.70, the unforced linearised system obeys:

$$\sum_{i=1}^{p}\sum_{k=0}^{r_i}\lambda_{ik}\frac{d^k y_i}{dt^k}=0 \quad (3.107)$$

for bounded initial conditions. We now want to investigate the conditions under which the internal dynamics are bounded for given initial conditions when the external input v is zero.

Internal stability. The following proposition, expanded from a previous SISO approach [42, 67], states that an implication of the global exponential stability of the zero dynamics is the internal stability of the system (boundedness of the solution $x(t, x(0))$ of system in Equation 3.2 under the control law in Equation 3.70) when the external input $v(t)$ is zero.

Proposition 3.1.5. *Consider the dynamic system given by Equation 3.2:*

3. Introduction to Feedback Linearisation

$$\dot{x} = f(x) + g(x)u$$
$$y = h(x) \tag{3.108}$$

with $x \in \Re^n$, $u \in \Re^p$ and $y \in \Re^p$, which has a vector relative degree $\{r_1, \cdots, r_p\}$, where $r = r_1 + \cdots + r_p < n$, and a normal form given by Equations 3.35 and 3.37. If (i) $\eta = 0$ is a globally exponentially stable equilibrium point of the zero dynamics of system in Equation 3.108, which are given by:

$$\dot{\eta}(t) = q_0(0, \eta(t)) \tag{3.109}$$

(ii) the mapping $q_0(\xi, \eta)$ is Lipschitz continuous in ξ,

$$||q_0(\xi_1, \eta) - q_0(\xi_2, \eta)|| \leq L_{q_0} ||\xi_1 - \xi_2|| \tag{3.110}$$

for all ξ_1 and ξ_2, and (iii) if the parameters λ_{ik} of the control law

$$u(t) = P(x) = -\left(\begin{bmatrix} \lambda_{1r_1} & \cdots & \lambda_{pr_p} \end{bmatrix} C(x) \right)^{-1} \left(\sum_{i=1}^{p} \sum_{k=0}^{r_i} \lambda_{ik} L_f^k h_i(x) \right) \tag{3.111}$$

are such that $\xi(t)$ is bounded for all t when the control in Equation 3.111 acts on system shown in Equation 3.108:

$$||\xi(t)|| \leq k_1 \tag{3.112}$$

then the solution $x(t, x(0))$ of system in Equation 3.108 under the control law in Equation 3.111:

$$\dot{x} = f(x) + g(x)P(x) = \bar{f}(x); \quad x(0) = x_o \tag{3.113}$$

is also bounded for all values of t, such that:

$$||x(t, x(0))|| \leq k_x \tag{3.114}$$

for some k_x, where $0 \leq k_x < +\infty$.

Proof. Using the converse Lyapunov Theorem 2.2.3, there exist a Lyapunov function V and positive constants such that

$$c_1 ||\eta||^2 \leq V(\eta) \leq c_2 ||\eta||^2 \tag{3.115a}$$

$$\frac{\partial V}{\partial \eta} q_0(0, \eta) \leq -c_3 ||\eta||^2 \tag{3.115b}$$

$$\left\| \frac{\partial V}{\partial \eta} \right\| \leq c_4 ||\eta|| \tag{3.115c}$$

Differentiating V with respect to time,

$$\dot{V} = \frac{\partial V}{\partial \eta} \dot{\eta} = \frac{\partial V}{\partial \eta} q_0(\xi, \eta) \tag{3.116}$$

3.1 Nonlinear Control Affine Systems

and adding each side of Equation 3.116 to Equation 3.115b:

$$\frac{\partial V}{\partial \eta} q_0(0, \eta) + \dot{V} \leq -c_3 ||\eta||^2 + \frac{\partial V}{\partial \eta} q_0(\xi, \eta) \tag{3.117}$$

the following expression is obtained:

$$\dot{V} \leq -c_3 ||\eta||^2 + \frac{\partial V}{\partial \eta} [q_0(\xi, \eta) - q_0(0, \eta)] \tag{3.118}$$

At this point, we cannot judge whether the second term on the right-hand side of Equation 3.118 helps or otherwise in making $\dot{V}(\eta)$ negative definite. The worst case would be that the contribution of this term is positive and equal to the product of two vector norms, as follows:

$$\dot{V} \leq -c_3 ||\eta||^2 + \left\| \frac{\partial V}{\partial \eta} \right\| ||q_0(\xi, \eta) - q_0(0, \eta)|| \tag{3.119}$$

Using Equation 3.115c and replacing the Lipschitz constant in Equation 3.110 into Equation 3.119, we obtain

$$\dot{V} \leq -c_3 ||\eta||^2 + c_4 ||\eta|| L_{q_0} ||\xi|| \tag{3.120}$$

Substituting the bound the norm of vector ξ given by Equation 3.112, gives

$$\dot{V} \leq -c_3 ||\eta||^2 + c_4 L_{q_0} k_1 ||\eta|| \tag{3.121}$$

Finally the Lyapunov function derivative satisfies $\dot{V} \leq 0$ for

$$||\eta|| \geq \frac{c_4 L_{q_0} k_1}{c_3} \tag{3.122}$$

which together with the bounds on $V(\eta)$ given by Equation 3.115a shows that

$$||\eta|| \leq \sqrt{\frac{c_2}{c_1}} \left(\frac{c_4 L_{q_0} k_1}{c_3} \right) = k_2 \tag{3.123}$$

Then the η variables are bounded. Since the ξ variables are also bounded and the transformation Φ is smooth, it follows that the solution $x(t, x(0))$ of the system given by Equation 3.108 under the control law given by Equation 3.111:

$$\dot{x} = f(x) + g(x)P(x) = \bar{f}(x); \quad x(0) = x_o \tag{3.124}$$

is also bounded for all t, such that

$$||x(t, x(0))|| \leq k_x \tag{3.125}$$

for some k_x, where $0 \leq k_x < +\infty$. $\square$

In the SISO case, Isidori [5] provides the following interesting proposition, which implies that the asymptotic stability of the equilibrium $\eta = 0$ of the zero dynamics, implies the asymptotic stability of the normal form equilibrium $(\xi, \eta) = (0, 0)$ under the feedback law in Equation 3.81 with $v = 0$:

Proposition 3.1.6. *Suppose that the equilibrium $\eta = 0$ of the zero dynamics of the SISO system given by Equation 3.44 is locally asymptotically stable and all eigenvalues of matrix A defined by Equation 3.84 have negative real parts, then the feedback law in Equation 3.81 with $v = 0$ provides an equilibrium $(\xi, \eta) = (0, 0)$ that is locally asymptotically stable.*

Proof. See [5].

Remark 3.1.5. Notice that the parameters $\lambda_0, \lambda_1, \cdots, \lambda_{r-1}$ can always be chosen to ensure that all eigenvalues of matrix A have negative real parts.

The conditions stated in the previous propositions are not necessary for local asymptotic stability of the feedback linearised system [68], as they are sufficient conditions. Nevertheless, in practice, the instability of internal dynamics would imply undesirable phenomena such as high temperatures, burning-up of fuses or harmful mechanical vibration of plant elements. For implementation purposes, the feasibility and effectiveness of a feedback linearising control law hinges upon and is guaranteed by stable internal dynamics.

We will now expand the notion of internal dynamic stability for zero external input $v = 0$, to the notion of the internal stability of the feedback linearised system when $v \neq 0$. With this purpose in mind, consider the following proposition:

Proposition 3.1.7. *Consider the system*

$$\dot{x} = f(x) + g(x)u \tag{3.126}$$

and suppose that the origin $x = 0$ is an asymptotically stable equilibrium point of the autonomous system $\dot{x} = f(x)$. Then, for all $\varepsilon > 0$ there exist $\delta_1 > 0$ and $K > 0$ such that if $||x(0)|| < \delta_1$ and $||u(t)|| < K$ for all $t > 0$, then the solution $x(t, x(0))$ of Equation 3.126 satisfies:

$$||x(t, x(0))|| < \varepsilon, \text{ for all } t \geq 0 \tag{3.127}$$

Proof. The proof of this proposition can be found in p. 516 of [5].

Using Proposition 3.1.7 a precise characterisation of the internal stability of the closed loop system is stated in the following theorem [57, 69].

Theorem 3.1.3. *The solution $x(t, x(0)), t > 0$, of system in Equation 3.2 under the control law given by Equation 3.75 with $P(x)$ and $Q(x)$ given by*

3.1 Nonlinear Control Affine Systems 53

Equations 3.76 and 3.77, respectively, is bounded if (i) the equilibrium point $x = 0$ of the autonomous system

$$\dot{x} = \bar{f}(x) = f(x) + g(x)P(x) \tag{3.128}$$

is asymptotically stable and (ii) the external input $v(t)$ and initial conditions $x(0)$ are such that for each $\epsilon > 0$, there exist δ_1 and $K > 0$, such that

$$||x(t, x(0))|| < \varepsilon \tag{3.129}$$
$$||x(0)|| < \delta_1 \tag{3.130}$$
$$||v(t)|| < K \tag{3.131}$$

Proof. Substituting u, given by the control law in Equation 3.75, into Equation 3.2, gives

$$\begin{aligned}\dot{x} &= f(x) + g(x)[P(x) + Q(x)v(t)] \\ &= f(x) + g(x)P(x) + g(x)Q(x)v(t) \\ &= \bar{f}(x) + \bar{g}(x)v(t)\end{aligned} \tag{3.132}$$

where $\bar{f}(x) = f(x) + g(x)P(x)$ and $\bar{g}(x) = g(x)Q(x)$. Based on Proposition 3.1.7, the stability of the internal dynamics for the autonomous system

$$\dot{x} = \bar{f}(x) \tag{3.133}$$

guarantees the boundedness of the solution $x(t, x(0))$, $t > 0$, of the non-autonomous system:

$$\dot{x} = \bar{f}(x) + \bar{g}(x)v(t) \tag{3.134}$$

such that for $\varepsilon > 0$ there exists $\delta_1 > 0$ and $K > 0$ that satisfy:

$$||x(t, x(0))|| < \varepsilon \tag{3.135}$$
$$||x(0)|| < \delta_1 \tag{3.136}$$
$$||v(t)|| < K \tag{3.137}$$

□

In the SISO case, if the equilibrium of the zero dynamics is asymptotically stable, we can ensure the asymptotic stability of the equilibrium of the autonomous system ($v = 0$) by appropriate choice of the parameters $\lambda_0, \cdots, \lambda_{r-1}$ (see Proposition 3.1.6). Then, according to Theorem 3.1.3, the equilibrium point of the non-autonomous input–output linearised system ($v \neq 0$) is guaranteed to be bounded. In the MIMO case, it is not difficult to extend this argument for the case of input–output linearisation and decoupling presented above (see Chapter 7 of Isidori [5]).

3.2 Review of Other Linearisation Techniques

This section briefly introduces other linearising techniques with conceptual similarities to the input–output linearisation problem described in the previous sections. The techniques are intended for control affine systems described by Equation 3.2.

Initially, the exact linearisation of affine systems is discussed as a particular case of the input–output linearisation problem [5]. This is followed by the notion of linearisation by immersion [70] and finally by the Volterra linearisation problem [71]. Although these approaches are mathematically structured in a different way, they preserve the same type of input–output behaviour as the approach described in Sections 3.1.3 and 3.1.4.

The following sections provide a brief review of the above techniques and a comparison to the input–output linearisation approach on which this book is based.

3.2.1 Exact linearisation

The exact linearisation via feedback consists of transforming an affine system described by Equation 3.2 under a static state feedback and a coordinate transformation into a controllable linear system

$$\dot{z} = Ax + Bv \tag{3.138}$$

A system of the form given in Equation 3.2 is state space exact linearisable if and only if there exists a relative degree $\{r_1, \ldots, r_p\}$ at x_0, a nonsingular characteristic matrix given by Equation 3.11 and the total relative degree is complete, such that $r_1 + \cdots + r_p = n$.

If this is the case, the set of functions

$$\xi_k^i = \phi_k^i(x) = L_f^{k-1} h_i(x), \ i = 1, \ldots, p, \ k = 1, \ldots, r_i \tag{3.139}$$

defines a coordinate transformation at x_0 such that,

$$\Phi(x) = \left[\phi_1^1(x), \ldots, \phi_{r_1-1}^1(x), \ldots, \phi_1^p(x), \ldots, \phi_{r_p-1}^p(x), \phi_{r+1}(x), \ldots, \phi_n(x) \right]^T \tag{3.140}$$

The system in the new coordinates is given by a set of differential equations,

$$\left. \begin{array}{l} \frac{d\xi_1^i}{dt} = d\xi_2^i(t) \\ \vdots \\ \frac{d\xi_{r_i-1}^i}{dt} = \xi_{r_i}^i(t) \\ \frac{d\xi_{r_i}^i}{dt} = b_i(\xi) + \sum_{j=1}^{p} c_{ij}(\xi) u_j \end{array} \right\}, \ i = 1, \ldots, p \tag{3.141}$$

3.2 Review of Other Linearisation Techniques

where $b_i(\cdot)$ is given by the ith element of the vector

$$b(x) = \begin{bmatrix} L_f^{r_1} h_1(x) \\ \vdots \\ L_f^{r_p} h_p(x) \end{bmatrix} \tag{3.142}$$

and $c_{ij}(\cdot)$ is the ijth element of the characteristic matrix given by Equation 3.11. Solving

$$b_i(\xi) + \sum_{j=1}^{p} c_{ij}(\xi) u_j = v_i \tag{3.143}$$

for each output guarantees exact linear dynamics for all the states when the input vector u has, in terms of the original description of the system, the form

$$u = C(x)^{-1}[-b(x) + v] \tag{3.144}$$

The linearised system in Equation 3.138 is finally given by

$$\left[\frac{\partial \Phi}{\partial x} \left(f(x) + g(x) \left(-C(x)^{-1} b(x) \right) \right) \right]_{x = \Phi^{-1}(z)} = Az \tag{3.145a}$$

$$\left[\frac{\partial \Phi}{\partial x} \left(g(x) C(x)^{-1} \right) \right]_{x = \Phi^{-1}(z)} = B \tag{3.145b}$$

The case of exact linearisation may be considered as a special case of the input–output linearisation. The condition for the problem to be solvable is the complete relative degree of the system $r = r_1 + \cdots + r_p = n$.

The main advantage is that the resulting system given by Equation 3.138 is favoured by all the stability properties of linear control theory. The asymptotic stability is given only by the position of the poles, without considering the internal or zero dynamics. Exact linearisation has proved to be a powerful tool in control of nonlinear systems. However, a number of physical systems cannot satisfy the restrictive conditions for applicability. The necessity of a complete relative degree $r = n$ dramatically restricts the subclass of systems of the form shown in Equation 3.2 that can be exactly linearised. In addition, the control law does not allow direct positioning of the closed loop poles, leading to possible inapplicable control signals to achieve the fixed linearisation.

3.2.2 Linearisation by immersion

A nonlinear system of the form shown in Equation 3.2 has linear input–output behaviour if it is immersed into a linear system of the following form

$$z = A_{n \times n} z + B_{n \times p} u$$
$$y = C_{p \times n} z \quad (3.146)$$

if there exists a transformation Ω such that for every $x^0 \in \Re^n$, the systems in Equations 3.2 and 3.146, both initialised at zero, have the same generating series [70]. In the original form, a very limited class of systems of the form 3.2 can be linearised by immersion at first. The problem of linearisation by immersion lies in finding a static state feedback of the form $u = P(x) + Q(x)v$, which is given by Equation 3.63, such that the resulting system can be immersed into a linear system of the form given in Equation 3.146. The conditions for immersability characterise a subclass of systems with nonsingular characteristic matrix, as is the case with the input–output linearisation. However, the control law is significantly more complicated, and demands more computational resources in its implementation.

3.2.3 Volterra linearisation

The problem of Volterra linearisation is finding a static feedback of the form given in Equation 3.95 that closes the loop on system 3.2, making it linear in the sense of the Volterra series, i.e. the input-dependent part of the Volterra series expansion and the one of an autonomous linear system are identical [71].

Taking a sequence of Toepliz matrices

$$\theta_k(x) = \begin{bmatrix} T_0(x) & T_1(x) & \cdots & T_k(x) \\ 0 & T_0(x) & \cdots & T_{k-1}(x) \\ \vdots & \vdots & \ddots & \vdots \\ 0 & 0 & \cdots & T_0(x) \end{bmatrix}, \quad k = 0, 1, \ldots \quad (3.147)$$

where

$$T_k(x) = L_g \begin{bmatrix} L_f^k h_1(x) \\ \vdots \\ L_f^k h_p(x) \end{bmatrix}, \quad k = 0, 1, \ldots, \quad (3.148)$$

denoting the rank of $\theta_k(x)$ for a fixed x by $\sigma[\theta_k(x)]$ and $\rho[\theta_k(x)]$ its rank as a matrix function, the system in Equation 3.2 is Volterra linearisable [71] under static state feedback if

$$\sigma[\theta_k(x)] = \rho[\theta_k(x)], \quad \text{for all } k \geqslant 0 \quad (3.149)$$

The concept of Volterra linearisation is more general than the input–output linearisation problem solved in Section 3.1.3. However, checking the rank conditions and finding the control law requires the use of Silverman's structure algorithm on the Toepliz matrices, compromising any practical implementation under restricted computational resources.

3.3 General Nonlinear Systems

The input–output linearisation techniques presented so far are founded on control affine systems of the form $\dot{x} = f(x) + g(x)u$. The input linearity of this kind of system enables the straightforward production of a feedback linearising control law that that makes the input–output behaviour linear. However, the mentioned techniques are applicable only to systems that are linear in the input variables. A more general class of dynamic systems in which the manipulated inputs may appear nonlinearly has the form

$$\dot{x} = f(x, u)$$
$$y = h(x) \tag{3.150}$$

In general, the solution to the input–output linearisation law for a system described by Equation 3.150 cannot be found analytically. However, two approaches have been proposed for an approximate input–output linearisation of nonaffine systems. These techniques, as well as more general concepts of Lie derivatives and relative degree, are briefly described and referenced below.

3.3.1 Lie derivative and relative degree

As with control affine systems, the following definitions specify the differential operation and the relative degree, respectively, for general nonlinear systems.

Definition 3.3.1. *The Lie derivative of a scalar function $\lambda(x)$ with respect to the vector function $f(x, u)$ is defined as*

$$[L_f \lambda](x, u) = \frac{\partial \lambda(x)}{\partial x} f(x, u) \tag{3.151}$$

Higher order Lie derivatives are defined as

$$[L_f^k \lambda](x, u) = \frac{[\partial L_f^{k-1} \lambda](x, u)}{\partial x} f(x, u) , \quad [L_f^0 h](x, u) = h(x) \tag{3.152}$$

Definition 3.3.2. *Given a multivariable nonaffine system as defined in Equation 3.150, the ith output y_i is said to have relative degree r_i at a point (x_0, u_0) if*

$$\frac{\partial}{\partial u}[L_f^k \lambda_i](x, u) = 0, \ i = 1, \ldots, p, \ k < r_i \tag{3.153}$$

and

$$\frac{\partial}{\partial u}[L_f^{r_i} \lambda_i](x, u) \neq 0 \tag{3.154}$$

at least for one output $i = 1, \ldots, p$. Accordingly, the vector relative degree of the system is defined as $\{r_1, r_2, \ldots, r_p\}$. It follows that the first r_i Lie derivatives do not depend explicitly on the input such that

$$[L_f^k \lambda_i](x, u) = L_f^k \lambda(x) , \quad 0 \leqslant k \leqslant r_i - 1 \tag{3.155}$$

3.3.2 Approximate input–output linearisation

Essentially, two approaches have been proposed to solve the problem of feedback linearisation of a system described by Equation 3.150. The first approach [72] numerically solves, when possible, the equation

$$\frac{d^r y(t)}{dt^r} = [L_f^r h](x, u) = v \tag{3.156}$$

The need for a suitable model of the system or the lack of an analytical solution imposes the use of search routines to find a local solution, which may result in an unsatisfactory result. The second design approach [73] is based on the development of an extended model that is directly coupled to a new manipulated input w, designed to be the time derivative of u.

$$w \equiv \dot{u} \tag{3.157}$$

The system in Equation 3.150 can then be written as an extended system which is in control affine form:

$$\begin{aligned} \dot{\hat{x}} &= \hat{f}(\hat{x}) + gw \\ y &= h(\hat{x}) \end{aligned} \tag{3.158}$$

where

$$\hat{x} = \begin{bmatrix} x \\ u \end{bmatrix}, \quad \hat{f}(\hat{x}) = \begin{bmatrix} f(x, u) \\ 0 \end{bmatrix}, \quad g = \begin{bmatrix} 0 \\ 1 \end{bmatrix} \tag{3.159}$$

Subsequently, the system is input–output linearised by a static state feedback as shown in Section 3.1 for affine systems. This technique imposes as a condition the differentiability of the input signal with respect to time [8].

3.4 Symbolic Algebra Software

The use of symbolic algebra packages such as MATHEMATICA® or MAPLE® may be an important aid when it comes to computing the relative degree, normal form, zero dynamics and feedback linearisation of a nonlinear system. For these purposes, de Jager [74] has developed a symbolic tool known as NONLINCON that works under MAPLE®, while Kwatny and Blankenship [75] have developed a set of symbolic routines for use within MATHEMATICA®.

3.5 Remarks

For notational simplicity the feedback linearisation techniques presented in this Chapter have been restricted to square systems having p inputs and

p outputs. The results established in this chapter are straightforwardly extended to systems having a different number of inputs and outputs $m \neq p$. The nonsingularity condition for the characteristic matrix in Equation 3.11 is replaced by the assumption of its full rank.

In effect, under the full rank assumption, Proposition 3.1.2 is still valid making the coordinate transformations of Theorem 3.1.1 suitable to deduce the linearising law. For non-square systems, the decoupling case is not applicable as described in this chapter. Decoupling certain outputs from a given input becomes a very specific case and is not considered in this work. In the majority of industrial multivariable processes, the final plant scheme to control has the same number of inputs and outputs. In their original state, processes are represented by a large number of variables and preliminary control procedures deal not only with the selection of relevant variables but also with the discrimination of inputs and outputs by following specific guidelines. The final abstraction of the process is generally a square plant on which decoupling techniques can be applied. The two feedback linearisation approaches presented in this section allow a system, in which the input variable may appear nonlinearly, to be approximately linearised. They overcome the important restriction of control affinity of the approach shown in Section 3.1, and other approaches with similar foundations [7, 11, 8, 63, 9]. Nevertheless, they all still have the critical drawback of requiring not only a mathematical model of the plant, but also a fully measured state vector. The approach presented in this book uses a dynamic neural network as an observer, based on which a linearising-decoupling feedback of the form presented in this chapter is synthesised and then applied to the system. Chapter 6 presents the feedback linearisation technique based on the identification procedure that is described in Chapter 4.

3.6 Summary

The idea of feedback linearisation lies in the transformation of nonlinear dynamics into a linear form by using state feedback. The complete linearisation of the system corresponds to input–state linearisation while a less restrictive and more application relevant input–output linearisation refers only to external linear conditions disregarding internal dynamics. When a multivariable system is input–output linearised and decoupled, it admits standard linear controller design, which facilitates the stabilisation and tracking control problems. This chapter has introduced a number of established notions for the analysis of nonlinear systems, such as the concepts of relative degree, normal form and zero dynamics. The input–output linearisation approach has been studied in detail for MIMO and SISO systems, and the stability properties of the resulting linearised systems have been analysed. In addition, other feedback linearisation techniques have been reviewed. The applicability of linearising-decoupling techniques depends on the availability of a mathematical model of the plant and a fully measurable state vector, making these requirements important drawbacks to practical implementations. Chapter 6 describes a combination of a conventional linearisation-decoupling technique presented here and a neural network based identification procedure from Chapter 4 to yield an approximate feedback linearising-decoupling technique for multivariable systems.

CHAPTER 4
DYNAMIC NEURAL NETWORKS

4.1 Introduction

From the quest for artificial intelligence by emulating the brain and behaviour of living organisms, there have been several encouraging outcomes. The study of the mechanics and structure of the brain has led to the development of new computational models, inspired by the connectionism of the nervous system, for solving complex numerical problems such as pattern recognition, system modelling and information processing. These mathematical models are known as artificial neural networks (ANNs) or simply neural networks (NNs) and they consist of a set of interconnected processing units or artificial neurons [76, 14]. Depending on the internal structure of the neuron and the interconnections between them, the neural network can be static or dynamic.

This book concentrates on the use of neural networks for system identification and control. It is thus relevant at this point to mention the most important features of neural networks for the purposes of system identification and control:

Nonlinear function approximation. Neural networks are able to approximate nonlinear maps between an input space and an output space [77]. Typically, a set of points taken from the input and output spaces, known as the *training set*, is employed by a *training algorithm* to adjust the neural network parameters. This training process enables the neural network to model the input–output map.

Dynamic system approximation. By incorporating dynamics in the processing units, neural networks are able not only to approximate static maps but also orbits and general dynamic system trajectories [78, 79].

Generalisation capabilities. Neural networks are capable of generalising when the inputs presented to the network have not been used during the training process. In the case of static networks the network produces a point on the surface it has reconstructed from training, while a dynamic network generates a trajectory for given initial conditions and input signals.

Parallel-distributed processing. The inherent parallel structure of neural networks makes them highly tolerant to faults and suitable for hardware implementations on VLSI circuits.

Vectorisation. Neural networks are able to model systems with multiple inputs and multiple outputs.

This chapter presents a general description of single layer networks leading to a more comprehensive analysis of dynamic (recurrent) neural networks (DNNs), including stability and convergence conditions. Section 4.2 gives a brief historic review on the area while Section 4.3 introduces a general neural network structure based on an inherently dynamic viewpoint [76]. Section 4.4 briefly introduces the multilayer perceptron network, while Section 4.5 discusses to a greater extent dynamic neural networks. Section 4.6 discusses training and initialisation while Section 4.7 discusses validation, cross-validation and structure selection of dynamic neural networks. Section 4.8 presents a simple yet illustrative training example involving a single link manipulator model.

4.2 Origins of Neural Computation

In recent decades there has been an increasing interest in the study of the mechanism and structure of the brain and how they may be emulated. In the early 1940s the pioneers of the neural computation field studied the potential and capabilities of interconnecting several basic components based on the model of a neuron [15]. Others were concerned with the adaptation laws involved in neural systems [80]. Later work resulted in the *perceptron* model, an architecture that has subsequently received much attention [81]. This work was followed by a rigorous analysis of the perceptron, proving many properties and pointing out limitations of several models [82]. In the 1970s a study based on biological and psychological evidence proposed several architectures of nonlinear dynamic systems with novel characteristics [30] which inspired the development of a particular nonlinear dynamic structure to solve numerical problems such as optimisation [25]. In 1986, work carried out by researchers in parallel distributed processing yielded a series of results and algorithms that gave a strong impulse to the area and provided the catalyst for much of the subsequent research in the field [16]. Nowadays there are several well-defined NN architectures to tackle a wide variety of problems. For a thorough collection of network architectures see [83].

4.3 Single Layer Neural Network Structure

A neural network is defined by a collection of basic processing elements. Each processing element (which is also known as artificial neuron or neural unit), has a specific structure: a weighted summation of its inputs, a linear single-input single-output transfer function and a nonlinear static function. A neural unit is illustrated in in Figure 4.1 [76].

4.3 Single Layer Neural Network Structure

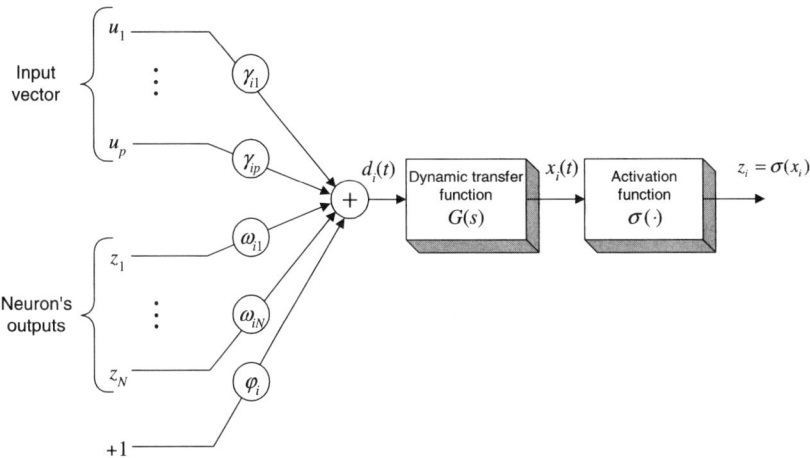

Fig. 4.1. Generic model of a neural unit

Weighted summation. The output from the weighted summation is given by

$$d_i(t) = \sum_{j=1}^{N} \omega_{ij} z_i(t) + \sum_{k=1}^{p} \gamma_{ik} u_k(t) + b_i \qquad (4.1)$$

with output elements z_i and inputs u_k, both weighted by scalars ω_{ij} and γ_{ik}, respectively. A bias b_i is also introduced to allow shifting. The vectorised form of an array of N of these units, which forms a single layer network, is given by

$$d(t) = \omega z(t) + \gamma u(t) + b \qquad (4.2)$$

where $\omega \in \Re^{N \times N}$, $\gamma \in \Re^{N \times p}$, $d = [d_1, \ldots, d_N]^T$, $z = [z_1, \ldots, z_N]^T$, $u = [u_1, \ldots, u_p]^T$ and $b = [b_1, \ldots, b_N]^T$.

Transfer function $G(s)$. The linear transfer function $G(s)$ is associated with input d_i and output x_i, such that

$$\bar{x}_i(s) = G(s) \bar{d}_i(s) \qquad (4.3)$$

where the bar denotes Laplace transformation of the variables. The most common choices for the transfer function $G(s)$ in neural networks are given in Table 4.1.

Activation function. The static nonlinear activation function $\sigma(\cdot)$ provides a mapping of the form:

$$z_i = \sigma(x_i) \qquad (4.4)$$

The choice of the activation function $\sigma(\cdot)$ depends on the intended application of the neural network. Activation functions can be classified as follows:

4. Dynamic Neural Networks

Table 4.1. Typical transfer functions used in neural networks

$G(s)$	Time domain dynamics
1	$x_i(t) = d_i(t)$
$\frac{1}{s}$	$\dot{x}_i(t) = d_i(t)$
$\frac{1}{s+\beta_i}$	$\dot{x}_i(t) + \beta_i x_i(t) = d_i(t)$
$\exp(-T_i s)$	$x_i(t) = d_i(t - T_i)$

- *Differentiable/non-differentiable.* Distinguishes smooth from sharp functions. Smooth functions are needed for some training algorithms such as backpropagation [16], whereas discontinuous functions are needed to give a true binary output.
- *Pulse-like/step-like.* Distinguishes functions that only have a significant output value for inputs near zero, from functions that only change significantly around zero.
- *Positive/zero-mean.* Refers to step-like functions. Positive functions change from 0 to 1 as the argument varies between $-\infty$ and $+\infty$, while zero-mean functions change from -1 instead of 0.

Some commonly used activation functions are given in Table 4.2. Notice that the sigmoid and hyperbolic tangent functions range from 0 to 1 and -1 to 1, respectively. Additionally, the threshold functions correspond to the high gain limits of the sigmoid and hyperbolic tangent functions. For the purposes of this book, the hyperbolic tangent function is a suitable choice. Properties such as smoothness, boundedness and monotony have been used many times in stability proofs in a crucial way. In addition, the simple form of its derivatives represents a useful resource when completing differential geometry structures.

The neurons by themselves are not very powerful in terms of computation or representation but their interconnections allows them to encode relations between the variables giving different processing capabilities. The three components of the neuron discussed above can be combined in a wide range of forms depending on the requirements of the application. The selection of a specific transfer function, activation function and interconnections defines a (single-layer) neural network architecture.

Example 4.3.1. Suppose that $G(s) = 1/s$. Each neural unit has the form

$$\dot{x}_i(t) = \sum_{j=1}^{N} w_{ij} z_j + \sum_{k=1}^{p} \gamma_{ik} u_k + b_i \qquad (4.5)$$

4.4 Static Multilayer Feedforward Networks

Table 4.2. Commonly used activation functions

Name	Formula	Features
threshold	1 if $x > 0$ else 0	non-differentiable, step-like, positive
threshold (bipolar)	1 if $x > 0$ else -1	non-differentiable, step-like, zero-mean
sigmoid	$(1 + \exp(-x))^{-1}$	differentiable, step-like, positive
hyperbolic tangent	$\tanh(x)$	differentiable, step-like, zero-mean
Gaussian	$\exp\left(-\frac{x^2}{\sigma^2}\right)$	differentiable, pulse-like
linear	x	differentiable

$$z_j = \sigma(x_j) \qquad (4.6)$$

while its corresponding vectorised form results in a single layer neural model of the form,

$$\begin{aligned} \dot{x} &= \omega z + \gamma u + b \\ z &= \sigma(x) \end{aligned} \qquad (4.7)$$

where $x = [x_1 \cdots x_N]^T$, $\sigma(x) = [\sigma(x_1) \cdots \sigma(x_N)]^T$, $z = [z_1 \cdots z_N]^T$, $u = [u_1 \cdots u_p]^T$, $b = [b_1 \cdots b_N]^T$. ∎

4.4 Static Multilayer Feedforward Networks

In the case of multilayer feedforward networks the connection of several layers gives the possibility of more complex nonlinear mapping between the inputs and the outputs. Such networks are typically static, having transfer function $G(s) = 1$. There is no feedback and only straight connections between consecutive layers. Therefore, the input of a layer is the output of the previous one (in the case of the first layer, the network input itself). For a static network with r layers,

$$\begin{aligned} x^{(1)} &= \gamma^{(1)} u + b^{(1)}; & z^{(1)} &= \sigma^{(1)}(x^{(1)}) \\ x^{(2)} &= \gamma^{(2)} z^{(1)} + b^{(2)}; & z^{(2)} &= \sigma^{(2)}(x^{(2)}) \\ &\vdots \\ x^{(r)} &= \gamma^{(r)} z^{(r-1)} + b^{(r)}; & z^{(r)} &= \sigma^{(3)}(x^{(r)}) \end{aligned}$$

There can be a different activation function at each layer. This type of neural network is widely used for nonlinear static modelling in many disciplines. An illustration of a three layer perceptron is given in Section 1.4.

4.5 Dynamic Neural Networks (DNNs)

The introduction of external feedback or internal dynamics into a feedforward neural network architecture produces a state space dynamic model. State feedback can be introduced, as will be shown below, by allowing the transfer function $G(s)$ to represent a dynamic system and not just a static gain. The general equation of a dynamic neural network is as follows:

$$\dot{x}(t) = f(x(t), u(t), \theta)$$
$$y(t) = g(x(t), \theta) \qquad (4.8)$$

where x are the states of the network, u the external input and θ is a vector of parameters of the network. The functions f and g are vector fields that define the dynamics and output mapping of the dynamic neural network, respectively.

Originally, recurrent networks were introduced in the context of associative or content addressable memory (CAM) problems [29] and [84]. The uncorrupted pattern is used as a stable equilibrium point and its noisy versions should lie in its basin of attraction. In this way, a dynamical system associated with a set of patterns is created. If the recurrent network correctly partitions the whole working space, then any initial condition (corresponding to a sample pattern) should have a steady-state solution corresponding to the uncorrupted pattern. The dynamics the recurrent neural network classifier serve as a filter.

The structure of the dynamic neural network mostly used in this book is a particular case of the general model given in Figure 4.1. The transfer function is given by the third row of Table 4.1:

$$G(s) = \frac{1}{s + \beta_i} \qquad (4.9)$$

such that each unit, which is denoted by the sub-index i, obeys the following differential equation:

$$\dot{x}_i = -\beta_i x_i + d_i \qquad (4.10)$$

Replacing d_i from Equation 4.1 into Equation 4.10, and assuming that no bias is used ($b_i = 0$), gives:

$$\dot{x}_i = -\beta_i x_i + \sum_{j=1}^{N} \omega_{ij} \sigma(x_j) + \sum_{j=1}^{p} \gamma_{ij} u_j \qquad (4.11)$$

where β_i, ω_{ij} and γ_{ij} are adjustable weights, $1/\beta_i$ is a positive time constant, x_i is the activation state of unit i, $u_1, \ldots, u_p$ are the input signals and $p \leq N$. This dynamic neural network is illustrated in Figure 4.2.

4.5 Dynamic Neural Networks (DNNs)

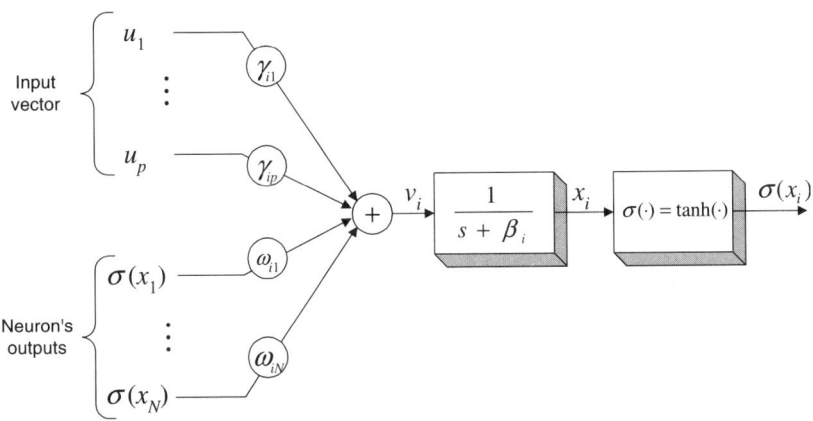

Fig. 4.2. Diagram of a dynamic neuron

A typical choice for the activation function is given by the fourth row of Table 4.2:

$$\sigma(x_j) = \tanh(x_j) \tag{4.12}$$

A dynamic neural network may be formed by a single layer of n units as in Equation 4.11. For convenience the output of the network will be taken as the first p states of vector x, leaving $(N-p)$ hidden states. The first p states of vector x are called output states. In this book, we will be mainly concerned with dynamic neural networks that have the same number of inputs and outputs. The vectorised form of equation Equation 4.11 is given by,

$$\dot{x} = -\beta x + \omega \sigma(x) + \gamma u$$
$$y_n = C_n x \tag{4.13}$$

where x are coordinates on $\Re^N$, $\omega \in \Re^{N \times N}$, $\sigma(x) = [\sigma(x_1), \ldots, \sigma(x_N)]^T$, $\gamma \in \Re^{N \times p}$, $u \in \Re^p$, $C_n = [I_{p \times N} 0_{p \times (N-p)}]$ and $\beta \in \Re^{N \times N}$ is a diagonal matrix with diagonal elements $\{\beta_1, \ldots, \beta_N\}$.

It is possible to make an analogy between the dynamic neural network described by Equation 4.13 with a general affine system with p inputs and p outputs of the form:

$$\dot{x} = f(x) + \sum_{j=1}^{p} g_j(x) u_j$$
$$y = h(x) \tag{4.14}$$

where $f(x), h(x), g_1(x), \ldots, g_p(x)$ are smooth vector fields defined on an open set of $\Re^n$. The dynamic neural network given by Equation 4.13 is a special case of the control affine system given by Equation 4.14, with

68   4. Dynamic Neural Networks

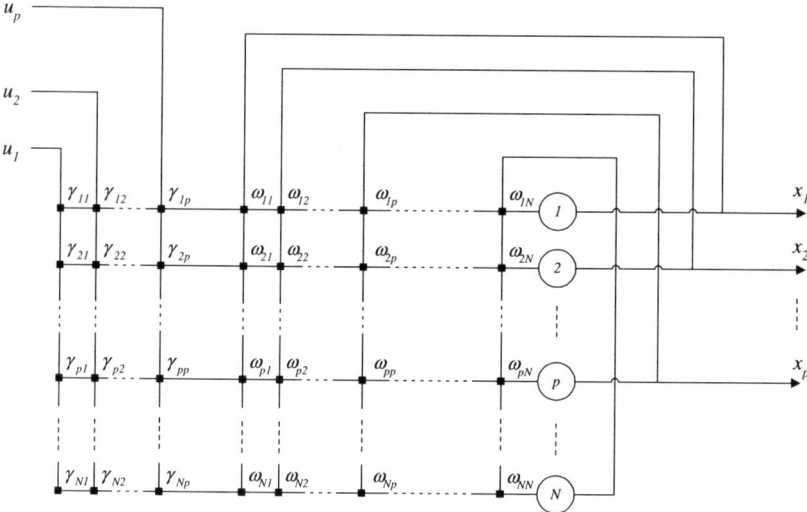

Fig. 4.3. Dynamic neural network with p output states and $n-p$ hidden units

$$f(x) = -\beta x + \omega \sigma(x)$$
$$g_j(x) = [\gamma_{1j}, \ldots, \gamma_{Nj}]^T$$
$$h(x) = C_n x$$

Note that hidden units are used to increase the dynamic mapping potential of the network. Their dynamics make it possible for networks to discover and exploit regularities of the task at hand, such as symmetries or replicated structure [85, 86].

4.5.1 Vector relative degree of a multi-input multi-output dynamic neural network

Given that the dynamic neural network described in Equation 4.13 is a control affine system of the form given in Equation 4.14, then a vector relative degree can be associated with the former. The characteristic matrix of the dynamic neural network is given by:

$$C(x) = \begin{bmatrix} L_{\gamma_1} L_f^{r_1-1} x_1 & L_{\gamma_2} L_f^{r_1-1} x_1 & \cdots & L_{\gamma_p} L_f^{r_1-1} x_1 \\ L_{\gamma_1} L_f^{r_2-1} x_2 & L_{\gamma_2} L_f^{r_2-1} x_2 & \cdots & L_{\gamma_p} L_f^{r_2-1} x_2 \\ \vdots & \vdots & \ddots & \vdots \\ L_{\gamma_1} L_f^{r_p-1} x_p & L_{\gamma_2} L_f^{r_p-1} x_p & \cdots & L_{\gamma_p} L_f^{r_p-1} x_p \end{bmatrix}_{p \times p} \quad (4.15)$$

which can be rewritten as:

$$C(x) = \begin{bmatrix} \sum_{i=1}^{N} \gamma_{i1} \frac{\partial L_f^{r_1-1} x_1}{\partial x_i} & \cdots & \sum_{i=1}^{N} \gamma_{ip} \frac{\partial L_f^{r_1-1} x_1}{\partial x_i} \\ \vdots & \ddots & \vdots \\ \sum_{i=1}^{N} \gamma_{i1} \frac{\partial L_f^{r_p-1} x_p}{\partial x_i} & \cdots & \sum_{i=1}^{N} \gamma_{ip} \frac{\partial L_f^{r_p-1} x_p}{\partial x_i} \end{bmatrix}_{p \times p} \quad (4.16)$$

and its vector relative degree is $\{r_1, r_2, \ldots, r_p\}$. The ith output y_i has relative degree r_i if

$$L_{\gamma_i} L_f^k x_i = 0, \, j = 1, \ldots, p, \, k = 0, \ldots, r_i - 2. \quad (4.17)$$

By definition, for all $k < r_i - 1$, the row vector

$$\begin{bmatrix} L_{\gamma_1} L_f^k x_1 & L_{\gamma_2} L_f^k x_2 & \ldots & L_{\gamma_p} L_f^k x_p \end{bmatrix} \quad (4.18)$$

is zero. For $k = r_i - 1$, this row vector is nonzero (i.e. has at least one non-zero element), because the characteristic matrix $C(x)$ is nonsingular by definition. Hence, for each i there is at least one choice of j such that the SISO system having output y_i and input u_j has exactly relative degree r_i and, furthermore, for any possible choice of j the corresponding relative degree is necessarily higher than or equal to r_i.

Each element r_i of the vector relative degree is associated with its corresponding output y_i. The structure of the dynamic neural network, which is defined by the values of N and p, together with the values of the parameters β, ω and γ, determine the vector relative degree associated with the dynamic neural network. Example 3.1.3 shows the calculation of the relative degree for a simple SISO dynamic neural network with two units. Example 3.1.4 provides the calculation of the vector relative degree for a two-input two-output dynamic neural network with four units. Reference [87] provides further details on the relative degree of dynamic neural networks.

4.5.2 Stability of dynamic neural networks

Dynamic neural network stability has been intensively studied since the late 1980s. First, local asymptotic stability was proved [88], showing that there could exist multiple equilibrium points, an aspect that is particularly useful for associative memory or pattern recognition dynamic networks. From the control systems point of view and the purposes of this book, global asymptotic stability (see Definition 2.2.4) and input-to-state stability are more relevant aspects to consider.

Consider a dynamic neural network of the form given by Equation 4.13. We will assume that the activation function $\sigma(x_i)$ has the following properties:

Assumption 4.5.1. $\sigma(x_i)$ is continuously differentiable.

Assumption 4.5.2. $0 \leq d\sigma/dx_i \leq 1$.

Assumption 4.5.3. $\sigma(x_i)$ is bounded, such that $|\sigma(x_i)| \leq \sup_z |\sigma(z)| < \infty$.

Notice that Assumptions 4.5.1 and 4.5.2 imply that $\sigma(x_i)$ is Lipschitz in x_i and also that the vector activation function is Lipschitz with constant L_σ:

$$\|\sigma(x) - \sigma(y)\| \leq L_\sigma \|x - y\| \text{ for all } x \in \Re^n, y \in \Re^n \tag{4.19}$$

From Table 4.2, the sigmoid function and the hyperbolic tangent function satisfy the above properties.

Proposition 4.5.1 indicates that under certain assumptions on the activation function, the DNN has at least one equilibrium point for each value of u.

Proposition 4.5.1. *If the vector activation function $\sigma(x)$ is continuous and satisfies $\|\sigma(x)\|_\infty \leq 1$ for all $x \in \Re^N$, then the dynamic neural network given by Equation 4.13 has at least one equilibrium point for each value of u, which corresponds to a fixed point of the mapping $\phi(x) = \beta^{-1}\omega\sigma(x) + \beta^{-1}\gamma u$.*

Proof. Assume that the network is known has an equilibrium point x for a constant input u. For the DNN described by Equation 4.13

$$\dot{x} = -\beta x + \omega \sigma(x) + \gamma u$$

an equilibrium point x satisfies the following:

$$x = \beta^{-1}\omega\sigma(x) + \beta^{-1}\gamma u$$

Define the map $\phi : \Re^N \to \Re^N$ as

$$\phi(x) = W\sigma(x) + S \tag{4.20}$$

with $W = \beta^{-1}\omega$ and $S = \beta^{-1}\gamma u$. Then, an equilibrium state x of the DNN given by Equation 4.13 is a fixed point of the map given by Equation 4.20, which satisfies:

$$\phi(x) = x \tag{4.21}$$

Let Ω be a hypercube defined by

$$\Omega = \{x : \|x - S\|_\infty \leq \|W\|_\infty\} \tag{4.22}$$

Then, $\|\phi(x) - S\|_\infty = \|W\sigma(x)\|_\infty \leq \|W\|_\infty \|\sigma(x)\|_\infty \leq \|W\|_\infty$ holds for all points in Ω since $\|\sigma(x)\|_\infty \leq 1$ by hypothesis. This implies that $\phi(x) \in \Omega$, so that ϕ is a continuous mapping from the bounded, convex, closed set Ω into itself. According to Brouwer's fixed point theorem [89], ϕ has at least one fixed point in Ω, thus the dynamic neural network given by Equation 4.13 has at least one equilibrium point for each value of u. □

4.5 Dynamic Neural Networks (DNNs)

It can easily be seen that the hyperbolic tangent activation function satisfies the hypothesis of Proposition 4.5.1.

Conditions for the uniqueness of the equilibrium point for dynamic neural networks (or Hopfield-type networks) have been studied by various authors from a static perspective. For instance, Guan and co-authors [90] provide some recent results. We will be more concerned, as will be seen below, with the conditions for global asymptotic stability, which, if satisfied, guarantee the uniqueness of the equilibrium point.

Let $x^* = [x_1^*, \ldots, x_N^*]^T$ be an equilibrium state of of the DNN described by Equation 4.13 for zero input, $u = 0$, such that

$$-\beta x^* + \omega\sigma(x^*) = 0 \qquad (4.23)$$

and consider the new variable,

$$\bar{x} = [\bar{x}_1, \ldots, \bar{x}_N]^T = x - x^* \qquad (4.24)$$

Differentiating Equation 4.24, using Equation 4.13 for $\dot{x}$, and given that $x = \bar{x} + x^*$, we obtain:

$$\begin{aligned}\dot{\bar{x}} &= \dot{x} - \dot{x}^* = \dot{x} \\ \dot{\bar{x}} &= -\beta(\bar{x} + x^*) + \omega\sigma(\bar{x} + x^*) + \gamma u\end{aligned} \qquad (4.25)$$

Adding and subtracting $\omega\sigma(x)$:

$$\dot{\bar{x}} = -\beta\bar{x} + \omega\left[\sigma(\bar{x} + x^*) - \sigma(x^*)\right] + \gamma u - \beta x^* + \omega\sigma(x^*) \qquad (4.26)$$

Using Equation 4.23, the last two terms of Equation 4.26 are zero, so that Equation 4.26 may be rewritten as follows:

$$\dot{\bar{x}} = -\beta\bar{x} + \omega\delta(\bar{x}) + \gamma u \qquad (4.27)$$

where $\delta(\bar{x})$ is a function given by:

$$\delta(\bar{x}) = [\delta_1(\bar{x}_1), \cdots, \delta_n(\bar{x}_n)]^T = \sigma(\bar{x} + x^*) - \sigma(x*) \qquad (4.28)$$

The following Lemma will be useful for later proofs.

Lemma 4.5.1. *Under Assumptions 4.5.1, 4.5.2 and 4.5.3, the vector function $\delta(x)$ satisfies:*

$$||\delta(\bar{x})|| \leq L_\sigma ||\bar{x}|| \qquad (4.29)$$

where L_σ is the Lipschitz constant of the vector activation function $\sigma(x)$.

Proof. It follows directly from the Lipschitz continuity of the vector activation function $\sigma(x)$. □

Two approaches are used in this chapter to study the stability of dynamic neural networks. The first approach uses a Lure-type Lyapunov function [91] to find sufficient conditions that ensure global asymptotic stability for the autonomous ($u = 0$) case. The second approach uses input-to-state stability analysis [92] for the non-autonomous case.

72   4. Dynamic Neural Networks

Stability by means of a Lure Lyapunov function.

Theorem 4.5.1. *The point $\bar{x} = 0$ is a globally asymptotically stable equilibrium point of the autonomous DNN given by:*

$$\dot{\bar{x}} = -\beta\bar{x} + \omega\delta(\bar{x}) \tag{4.30}$$

if the following sufficient conditions are satisfied:

1. *The activation function $\sigma(x_i)$ satisfies Assumptions 4.5.1, 4.5.2 and 4.5.3.*
2. *The matrix*

$$\hat{M} = -\beta + \frac{1}{2}(\omega + \omega^T) \tag{4.31}$$

is negative definite, i.e. all its eigenvalues are negative,

$$\lambda_{\max}(\hat{M}) < 0. \tag{4.32}$$

Proof. In order to find a sufficient condition that guarantees the origin $\bar{x} = 0$ to be a globally asymptotically stable equilibrium point of the system given by Equation 4.30, consider the following candidate Lure-type Lyapunov function,

$$V(\bar{x}) = \sum_i \int_0^{\bar{x}_i} \delta_i(z) dz \tag{4.33}$$

This kind of function is know as Lure-type Lyapunov function after the work by the Russian scientist A.I. Lure and due to its integral form [56]. From Equation 4.28, and Assumptions 5.3.1 to 5.3.3, it is easy to see that $V(\bar{x})$ is positive definite. Differentiating Equation 4.33 along trajectories of the system, the following is obtained,

$$\dot{V}(\bar{x}) = \sum_i \delta_i(\bar{x})\dot{\bar{x}}_i = \delta(\bar{x})^T\dot{\bar{x}} = \delta(\bar{x})^T[-\beta\bar{x} + \omega\delta(\bar{x})] \tag{4.34}$$

so that

$$\dot{V}(\bar{x}) = -\delta(\bar{x})^T\beta\bar{x} + \delta(\bar{x})^T\omega\delta(\bar{x}) \tag{4.35}$$

From Equation 4.28, and Assumptions 5.3.1 to 5.3.3, it follows that

$$\delta(\bar{x})^T\beta\delta(\bar{x}) \leq \delta(\bar{x})^T\beta\bar{x} \tag{4.36}$$

so that,

$$\dot{V}(\bar{x}) \leqslant \delta(\bar{x})^T(-\beta + \omega)\delta(\bar{x}) = -\delta(\bar{x})^T\beta\delta(\bar{x}) + \delta(\bar{x})^T\omega\delta(\bar{x}) \tag{4.37}$$

Matrix ω is in general not symmetric, but the quadratic form satisfies:

$$\delta(\bar{x})^T \omega \delta(\bar{x}) = \delta(\bar{x})^T \left[\frac{1}{2}(\omega + \omega^T)\right] \delta(\bar{x}) \tag{4.38}$$

so that

$$\dot{V}(\bar{x}) \leqslant \delta(\bar{x})^T \left[-\beta + \frac{1}{2}(\omega + \omega^T)\right] \delta(\bar{x}). \tag{4.39}$$

If $\hat{M}$ as given by Equation 4.31 is negative definite, then the right hand side of Equation 4.39 is negative. It follows that $\dot{V}(\bar{x})$ (Equation 4.35) is negative and therefore $V(\bar{x})$ (Equation 4.33) is a Lure-type Lyapunov function for the autonomous DNN given by Equation 4.30. Hence, the origin $\bar{x} = 0$ is a globally asymptotically stable (and hence unique) equilibrium point of the system. □

Notice that since β is diagonal and $(\omega + \omega^T)$ is symmetric, then $\hat{M}$ is symmetric and therefore all its eigenvalues are real. The condition provided by Theorem 4.5.1 [91] is less restrictive to the network parameters β and ω than other approaches [93], [94], [95]. Other conditions for the global asymptotic stability of dynamic neural networks are given by [96] and [97].

Example 4.5.1. To illustrate the categorisation of globally asymptotically stable networks, the following example is given. For an autonomous DNN described by,

$$\begin{bmatrix} \dot{x}_1 \\ \dot{x}_2 \\ \dot{x}_3 \end{bmatrix} = \begin{bmatrix} 0.7 & 0 & 0 \\ 0 & 1 & 0 \\ 0 & 0 & 0.8 \end{bmatrix} \begin{bmatrix} x_1 \\ x_2 \\ x_3 \end{bmatrix} + \begin{bmatrix} -0.3 & 0.3 & 1.1 \\ -1.4 & -1 & 0.2 \\ -0.7 & 0.3 & 0.2 \end{bmatrix} \begin{bmatrix} \tanh(x_1) \\ \tanh(x_2) \\ \tanh(x_3) \end{bmatrix} \tag{4.40}$$

the matrix $\hat{M}$ in Equation 4.31 and its eigenvalues are given by

$$\hat{M} = \begin{bmatrix} -1 & -0.55 & 0.2 \\ -0.55 & -2 & 0.25 \\ 0.2 & 0.25 & -0.6 \end{bmatrix} \tag{4.41}$$

$$\text{eig}(\hat{M}) = \begin{bmatrix} -0.7835 & -0.5167 & -2.2998 \end{bmatrix} \tag{4.42}$$

Condition 4.32 is satisfied and therefore the network is globally asymptotically stable. The phase plot given in Figure 4.4 shows different trajectories in the state space converging to the origin. ■

Input-to-state stability analysis for DNNs. Current approaches for nonlinear system stability can be divided into:

- The input–output approach, which relies on operator-theoretic techniques and requires the operator that represents the system to be bounded, such that bounded inputs map bounded outputs.

4. Dynamic Neural Networks

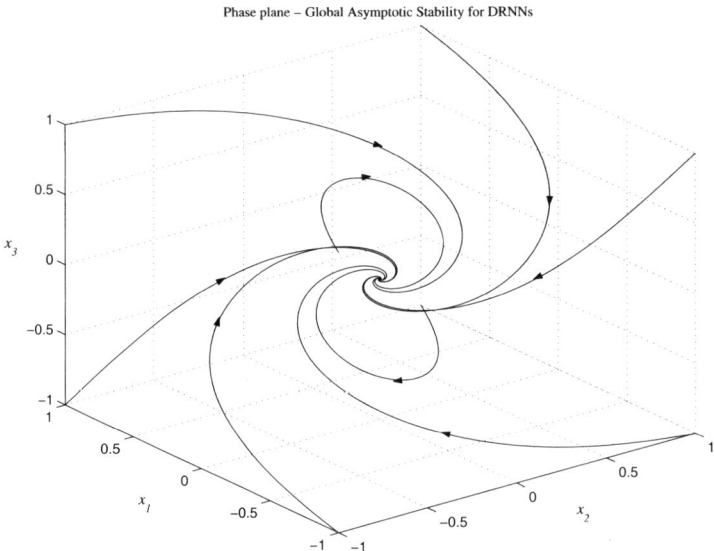

Fig. 4.4. Global asymptotic stability of a DNN

- The state space approach, where the basic object is the forced dynamic system represented by differential equations (or difference equations), and where the main tools are the stability theorems of Lyapunov.

These two paradigms are not equivalent for nonlinear systems, but the input-to-state stability notion [98] has allowed the links between the two approaches to be established and has become an important tool in nonlinear system analysis.

Definition 4.5.1 (Input-to-state stability). *A general nonlinear system*

$$\dot{x} = f(x, u) \tag{4.43}$$

with $f(\cdot, \cdot)$ locally Lipschitz, $f(0,0) = 0$, and the solution $x = 0$ of $\dot{x} = f_0(x) = f(x, 0)$ globally asymptotically stable, is input-to-state stable if
  1. *Its solution is continuous at $(0,0)$.*
  2. *There exists a class K function $\gamma_1(\cdot)$ such that*

$$\lim_{t \to +\infty} \sup \|x(x_0, u, t)\| \leqslant \gamma_1 \left(\|u\|_\infty\right) \tag{4.44}$$

uniformly for x_0 in any compact set and all u.

For a bounded and constant control input u, the state trajectory will be in a ball of radius $\gamma_1(\|u\|_\infty)$ when t is very large, regardless of the initial conditions. The uniformity requirement means that for each $r > 0$ and $\epsilon > 0$, there is a $T > 0$ such that $||x(x_o, u, t)|| < \epsilon + \gamma_1(||u||_\infty)$ for all u and all x_o with $||x_o|| < r$ and $t \geq T$.

Before introducing the theorems on input-to-state stability, it is relevant to state the following definition [98]:

Definition 4.5.2 (ISS–Lyapunov function). *A function $V : \Re^n \to \Re_+$, which is continuously differentiable, radially unbounded and positive definite, is an input-to-state (ISS) Lyapunov function, if it satisfies*

$$\dot{V}(x(t)) \leq \psi(||u(t)||) - \alpha(||x(t)||) \tag{4.45}$$

for all $x \in \Re^n$ and $u \in \Re^p$, with both $\alpha(\cdot)$ and $\psi(\cdot)$ class K_∞ functions.

The existence of an ISS–Lyapunov function is a sufficient condition for input-to-state stability. This result is also converse. In other words, the property of input-to-state stability implies the existence of an ISS-Lyapunov function. The following theorem summarises these facts:

Theorem 4.5.2. *A system described by Equation 4.43 is input-to-state stable if and only if there exists an ISS–Lyapunov function.*

Proof. See [99].

Theorem 4.5.3, which is inspired by the analysis given in [92], establishes sufficient conditions for input-to-state stability in the case of a dynamic neural network described by Equation 4.13.

Theorem 4.5.3. *The system in Equation 4.27 with an external input u different from 0, represented by,*

$$\dot{\bar{x}} = -\beta\bar{x} + \omega\delta(\bar{x}) + \gamma u \tag{4.46}$$

or, equivalently, the dynamic neural network given by Equation 4.13, is input-to-state stable if the following sufficient conditions are satisfied:

1. *The activation function $\sigma(x_i)$ satisfies Assumptions 4.5.1, 4.5.2 and 4.5.3.*
2. *Given $\mu \in \Re_+$, there exists a symmetric and positive definite solution P to the Lyapunov equation:*

$$-\beta^T P - P\beta = -\mu I \tag{4.47}$$

3. *The following inequalities are satisfied:*

$$||\omega||^2 < \frac{\mu - 2\,||P||}{||P||} \;,\; ||P|| < \frac{\mu}{2} \tag{4.48}$$

Proof. Consider the function $V(\bar{x})$ and its time derivative,

$$V(\bar{x}) = \bar{x}^T P \bar{x}, \ P = P^T > 0$$
$$\dot{V}(\bar{x}) = \bar{x}^T P \dot{\bar{x}} + \dot{\bar{x}}^T P \bar{x} \tag{4.49}$$

Substituting $\dot{\bar{x}}$ by Equation 4.46:

$$\dot{V}(\bar{x}) = \bar{x}^T P(-\beta \bar{x} + \omega \delta(\bar{x}) + \gamma u) + (-\beta \bar{x} + \omega \delta(\bar{x}) + \gamma u)^T P \bar{x}$$
$$= -\bar{x}^T (P\beta + \beta^T P) \bar{x} + \bar{x}^T P \omega \delta(\bar{x}) + \delta(\bar{x})^T \omega^T P \bar{x} + \bar{x}^T P \gamma u + u^T \gamma^T P \bar{x} \tag{4.50}$$

Using Equation 4.47, the following expression is obtained:

$$\dot{V}(\bar{x}) = -\mu \bar{x}^T \bar{x} + 2\bar{x}^T P \omega \delta(\bar{x}) + 2\bar{x}^T P \gamma u \tag{4.51}$$

Consider the identity:

$$X^T Y + Y^T X \leq X^T \Lambda X + Y^T \Lambda^{-1} Y \tag{4.52}$$

which is valid for any matrices $X \in \Re^{n \times k}$ and $Y \in \Re^{n \times k}$, with $\Lambda \in \Re^{n \times n}$ positive definite and symmetric.

For the second term of Equation 4.51, using the identity given in Equation 4.52 with $X = \bar{x}$, $Y = P\omega\delta(\bar{x})$ and $\Lambda = P$, we have

$$2P\omega\delta(\bar{x}) \leq x^T P x + (P\omega\delta(\bar{x}))^T P^{-1} P\omega\delta(\bar{x})$$
$$= x^T P x + \delta(\bar{x})^T \omega^T P^T \omega \delta(\bar{x}) \tag{4.53}$$
$$\leq ||P|| ||\bar{x}||^2 + ||P|| ||\omega||^2 ||\delta(\bar{x})||^2$$

but, if the activation function $\sigma(x_i)$ satisfies Assumptions 4.5.1, 4.5.2 and 4.5.3, then by Lemma 4.5.1 $||\delta(\bar{x})||^2 \leq L_\sigma^2 ||\bar{x}||^2$,

$$2P\omega\delta(\bar{x}) \leqslant ||P|| \, ||\bar{x}||^2 + ||P|| \, ||\omega||^2 L_\sigma \, ||\bar{x}||^2 \tag{4.54}$$

For the third term in Equation 4.51, using once again the identity in Equation 4.52, with $X = \bar{x}$, $Y = P\gamma u$ and $\Lambda = P$, we have

$$2\bar{x}^T P \gamma u \leq x^T P x + (P\gamma u)^T P^{-1} P \gamma u$$
$$= x^T P x + u^T \gamma^T P^T \gamma^T u \tag{4.55}$$

so that

$$2\bar{x}^T P \gamma u \leqslant ||P|| \, ||\bar{x}||^2 + ||P|| \, ||\gamma||^2 \, ||u||^2 \tag{4.56}$$

Substituting Equations 4.54 and 4.56 into 4.51

$$\dot{V}(\bar{x}) \leqslant (-\mu + 2\,\|P\| + L_\sigma^2\,\|P\|\,\|\omega\|^2)\,\|\bar{x}\|^2 + \|P\|\,\|\gamma\|^2\,\|u\|^2 \qquad (4.57)$$

Defining

$$\begin{aligned}\alpha(r) &= -(-\mu + 2\,\|P\| + L_\sigma^2\,\|P\|\,\|\omega\|^2)r^2 \\ \psi(r) &= \|P\|\,\|\gamma\|^2\,r^2\end{aligned} \qquad (4.58)$$

for $r \in \Re$, such that

$$\dot{V}(\bar{x}) \leqslant \psi(\|u\|) - \alpha(\|\bar{x}\|) \qquad (4.59)$$

the function $V(\bar{x})$ is an ISS-Lyapunov function if $\alpha(\cdot)$ and $\psi(\cdot)$ are K_∞ functions [56]. Since $\psi(\cdot)$ already satisfies this condition, the system in Equation 4.46 is input-to-state stable when

$$-\mu + 2\,\|P\| + L_\sigma^2\,\|P\|\,\|\omega\|^2 < 0 \qquad (4.60)$$

which implies the following,

$$\|\omega\|^2 < \frac{\mu - 2\,\|P\|}{L_\sigma^2\,\|P\|},\quad \|P\| < \frac{\mu}{2} \qquad (4.61)$$

□

These conditions are less restrictive than those imposed in other approaches that use diagonally arranged matrices [100, 101].

Example 4.5.2. Consider the dynamic neural network described by:

$$\begin{bmatrix}\dot{x}_1 \\ \dot{x}_2\end{bmatrix} = -\begin{bmatrix}2 & 0 \\ 0 & 2\end{bmatrix}\begin{bmatrix}x_1 \\ x_2\end{bmatrix} + \begin{bmatrix}0.3 & 0.8 \\ 0.4 & 0.3\end{bmatrix}\begin{bmatrix}\tanh(x_1) \\ \tanh(x_2)\end{bmatrix} + \begin{bmatrix}1 \\ 1\end{bmatrix}u \qquad (4.62)$$

Selecting $\mu = 1$ and solving the Lyapunov Equation 4.47 gives:

$$P = \begin{bmatrix}0.25 & 0 \\ 0 & 0.25\end{bmatrix} \qquad (4.63)$$

We have that the induced 2-norms are $\|P\| = 0.25$ and $\|\omega\| = 0.9606$. The Lipschitz constant of the vector activation function for the Euclidean norm is $L_\sigma = 1$. Therefore,

$$\frac{\mu - 2\|P\|}{\|P\|} = 2 > \|\omega\|^2 = (0.9606)^2 = 0.9227 \qquad (4.64)$$

and $\|P\| = 0.25 < \mu/2 = 0.5$; therefore by Theorem 4.5.3, this DNN is input-to-state stable. Figure 4.5 shows a simulation of the DNN response with a square wave input of magnitude 2 and frequency 0.1 Hz. Notice the stable response as expected. ∎

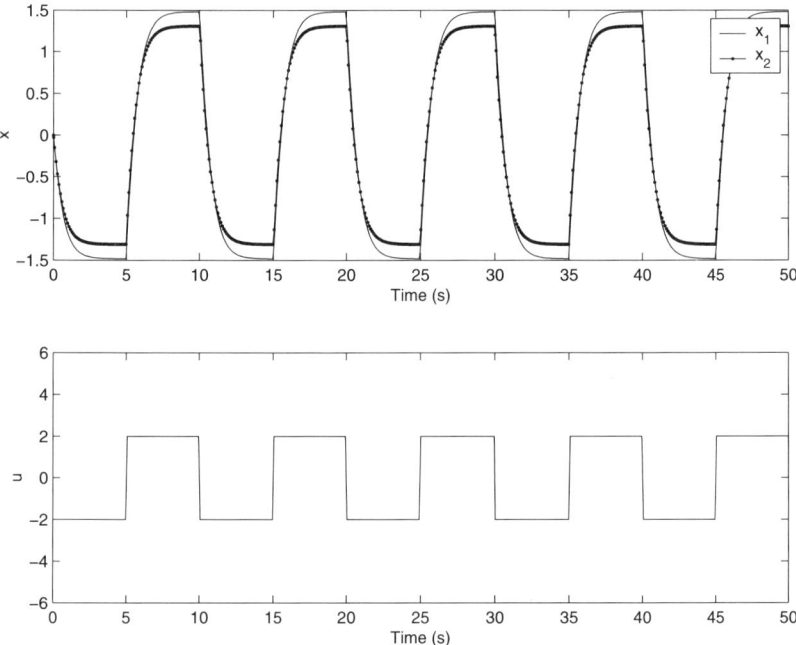

Fig. 4.5. Response of the DNN of Example 4.5.2 with a square wave input

4.6 Training Dynamic Neural Networks

4.6.1 Experiment design and input characterisation

Before an empirical model can be identified, a data set has to be collected from the system that the designer wishes to model. Assuming that the sensors and actuators are in place and that the designer knows what inputs and outputs are required in the model, important choices left are the character of the input, the sampling frequency and the length of the experiment.

To characterise the input, it is important to identify the operating range of the system, and the region of the operating range where the model is required. The designer should be careful not to excite dynamic modes (such as resonances) that are not required to be incorporated in the model.

Another important notion for choosing the character of the input is that of persistent excitation [31]. The input should be sufficiently variable so that the resulting data contain sufficient information to enable the estimation of the parameters in the model.

For the identification of linear models, it is very common to use the pseudo-random-binary-sequence (PRBS), which consists of random changes between two values, with a certain switching probability. However, the identification of nonlinear models requires the input to have more variability in terms of amplitude.

Some possible inputs for the identification of nonlinear systems are briefly described below. Further information on the characterisation of the input can be found, for instance, in References [31] and [102].

Sum of sinusoids. This type of input signal is given by:

$$u(t) = \sum_{j=1}^{M} a_j \sin(\omega_j t + \phi_j) \qquad (4.65)$$

where the angular frequencies ω_j, amplitudes a_j and phase shifts ϕ_j are different.

Discrete time white noise signal. This signal is characterised by a sequence $\{e(k)\}$ of independent, equally distributed random variables with a certain probability density function. If the sampling time is fast, this signal may cause the wear of actuators.

Constant input for N samples. This discrete time signal is given by:

$$u(k) = e\left(\text{int}\left[\frac{k-1}{N}\right] + 1\right) \qquad (4.66)$$

where k is an integer, e is a discrete time white noise process with zero mean and standard deviation σ_e, and int[ ] denotes the integer part of a real number.

Random steps. This is a discrete time signal where steps of random magnitude may occur at sampling instants with a certain probability α.

$$u(k) = \begin{cases} u(k-1) & \text{with probability } (1-\alpha) \\ e(k) & \text{with probability } \alpha \end{cases} \qquad (4.67)$$

where k is an integer, e is a discrete time white noise process with zero mean and standard deviation σ_e.

4.6.2 Training as an optimisation problem

Suppose that data has been collected from a real system that is to be modelled by means of a dynamic neural network described by Equation 4.8. Consider a training data set with M input–output pairs and sampling time T_s:

$$Z_M = [y(t_k), u(t_k)]_{k=1,M} \qquad (4.68)$$

where $y \in \Re^p$ is the measured output, $u \in \Re^m$ is the input variable, and k is a sampling index. The problem of training the DNN to learn the dynamics from the data set Z_M may be reduced to a conventional optimisation problem.

Given a model structure such as the one described by Equation 4.8, the Prediction Error Method [31] attempts to find the estimated parameter vector

$\theta \in \Re^{n_\theta}$ such that an objective function (typically the mean square error) is minimised:

$$F_M(\theta, Z_M) = \frac{1}{2M} \sum_{k=1}^{M} ||y(t_k) - \hat{y}(t_k|\theta)||^2 \qquad (4.69)$$

where $\hat{y}(t_k|\theta)$ is the output vector of the model at time t_k given the decision vector θ, and $||\cdot||$ denotes the Euclidean norm.

If the sampling time is short compared with the dynamics of the system that generated the data to be fitted, then the objective function F may be written as follows:

$$F_M(\theta, Z_M) = \frac{1}{2(t_M - t_1)} \int_{t_1}^{t_M} ||y(t) - \hat{y}(t|\theta)||^2 dt \qquad (4.70)$$

where the integral sign denotes numerical quadrature.

A dynamic neural network training problem can be cast as a nonlinear unconstrained optimisation problem:

$$\min_{\theta} \quad F_M(\theta, Z_M) \qquad (4.71)$$

The optimisation is typically carried out using unconstrained quasi-Newton methods, random search methods or genetic algorithms (GAs). The optimisation problem associated with training usually exhibits local minima, so several runs from different (possibly random) initial decision variables should be made, noting that genetic algorithms are less sensitive to the initial values of the decision variables than quasi-Newton methods, at the expense of higher computational requirements.

For training purposes, the decision vector that has typically been reported in the literature is based on the matrix coefficients of the DNN in Equation (4.13):

$$\theta = \begin{bmatrix} \beta_d \\ \text{vec}(\omega) \\ \text{vec}(\gamma) \end{bmatrix} \qquad (4.72)$$

where β_d is a vector with the diagonal elements of β and $\text{vec}(\cdot)$ is a vector created with the stacked columns of an argument matrix $(\cdot)$.

Notice that the state vector x of the DNN of the form given by Equation 4.13 can be partitioned into the output states x^o and the hidden states x^h:

$$x = \begin{bmatrix} x^o \\ x^h \end{bmatrix} \qquad (4.73)$$

where $x^o \in \Re^p$ and $x^h \in \Re^{n-p}$.

The decision vector θ may be augmented, as proposed in [103] to include the initial values of the hidden states of the DNN, $x^h(t_1)$:

4.6 Training Dynamic Neural Networks

$$\theta = \begin{bmatrix} \beta_d \\ \text{vec}(\omega) \\ \text{vec}(\gamma) \\ x^h(t_1) \end{bmatrix} \tag{4.74}$$

The training procedure would be as follows:

Procedure 4.6.1 (DNN Training).

- Step 1: Initialise the output states as follows:
$$x^o(t_1) = y(t_1) \tag{4.75}$$
- Step 2: Initialise the values of β_d, ω, γ and $x^h(t_1)$ with random values. Form the initial decision vector $\theta^{(0)}$ according to Equation (4.74).
- Step 3: Compute the decision vector $\hat{\theta}$ by solving the associated optimisation problem, with $F(\theta, Z_M)$ given by either Equation (4.69) or (4.70):
$$\hat{\theta} = \arg\min_{\theta} \; F_M(\theta, Z_M) \tag{4.76}$$

Notice that the output states are normally initialised with the values of the output variables. Also notice that several runs of Procedure 4.6.1 from different initial parameters may be required to check for local minima.

If the initial values of the hidden states are not included in the vector of decision variables, then it is likely that the resulting model parameters β, ω and γ will be biased. The explanation of this is as follows. Although the initial values of the hidden states $x^h(t_1)$ do not affect the initial values of the output states $x^o(t_1)$ (which are typically set to match the real system output), the initial values of the hidden states do affect the initial time derivatives of the output states $\dot{x}^o(t_1)$. This can easily be inferred by looking at the following form of Equation 4.13 at time t_1 and recalling that ω is a full matrix:

$$\frac{d}{dt}\begin{bmatrix} x^o(t_1) \\ x^h(t_1) \end{bmatrix} = -\beta \begin{bmatrix} x^o(t_1) \\ x^h(t_1) \end{bmatrix} + \omega\sigma\left(\begin{bmatrix} x^o(t_1) \\ x^h(t_1) \end{bmatrix}\right) + \gamma u(t_1) \tag{4.77}$$

So, by fixing the initial values of the hidden states to zero for training purposes, the initial time derivatives of the DNN outputs may have different values than the time derivatives of the data that the network is required to learn. This will force the training algorithm to adjust the network parameters β, ω and γ to compensate for the initial incorrect values in the derivatives, so introducing a bias in the values of these parameters.

In linear state space modelling, the effect of the initial states depends on the stability of the model and can be separated from the influence of the external input. If the linear model is asymptotically stable, then the effect of the initial states decays exponentially with time, and if the system dynamics are fast, then the effect of the initial states disappears quickly [31].

On the other hand, in nonlinear state space modelling, the effect of the initial states is richer and cannot in general be separated from the effect of the external input [69].

4.6.3 Initialising dynamic neural networks

Once a DNN has been trained, it is likely that the designer will use it on a different data set for validation or simulation purposes, or it may even be used on-line as part of a monitoring or nonlinear control scheme.

Denote $y_v(t_1)$ as the value of the output vector at time t_1 and $u_v(t_1)$ as the value of the input vector at time t_1.

A simple two step procedure can be used to initialise the DNN at time t_1

Procedure 4.6.2 (DNN Initialisation).

- Step 1: Initialise the output states $x^o(t_1)$ with the values of the system output at time t_1:

$$x^o(t_1) = y(t_1) \tag{4.78}$$

- Step 2: Initialise the hidden states $x^h(t_1)$ by solving the following optimisation problem:

$$x^h(t_1) = \arg\min_{x^h(t_1)} \|\dot{y}(t_1) - \dot{x}^o(t_1)\|^2 \tag{4.79}$$

where $\dot{y}(t_1)$ can be calculated from known data using finite differences and $\dot{x}^o(t_1)$ can be calculated from Equation 4.77, given the values of $x^o(t_1)$, $x^h(t_1)$, β, ω and γ.

If the DNN is being used for validation purposes, then $\dot{y}(t_1)$ may be calculated using the validation data set. On the other hand, if the DNN is being used in an on–line application, then $\dot{y}(t_1)$ can be calculated using the recent history of $y(t)$.

By using Procedure 4.6.2, it is ensured that the initial value of the hidden state vector $x^h(t_1)$ is such that the initial output derivative of the DNN, $\dot{x}^o(t_1)$, is as close as possible to the required value $\dot{y}(t_1)$.

Notice that in general the number of hidden states is different from the number of outputs. Therefore, formulating the DNN initialisation as an optimisation problem is more general than formulating it as a nonlinear equation solution problem.

4.6.4 Gradient based optimisation methods

We have seen above that training a dynamic neural network involves solving an unconstrained optimisation problem in several variables. Gradient based methods are well known optimisation methods that can be used for training neural networks and this subsection introduces their fundamentals. As their name suggests, gradient based methods require the calculation, or numerical approximation, of the derivatives of the objective function with respect to the parameters, so they require the objective function to be sufficiently smooth. Due to their local nature, these methods can only find a local minimum given the initial guesses for the decision variables. Two well known references for this topic are [104, 105]. Two gradient based methods that are also used for training recurrent neural networks but which are not treated in this book are the Levemberg–Marquardt algorithm [35], and the recurrent backpropagation method (also known as backpropagation through time) [14].

Consider the optimisation problem:

$$\min_{\theta \in \Re^n} F(\theta) \qquad (4.80)$$

The function $F : \Re^n \to \Re$ is a real valued function called the objective function. The vector θ is an n-vector of independent variables, that is: $\theta = [\theta_1, \theta_2, \cdots \theta_n]^T$. The variables $\theta_1, \theta_2, \cdots \theta_n$ are often referred to as *decision variables*.

Assume that function F has first and second partial derivatives, so that using a Taylor series the function can be approximated about $\bar{\theta}$:

$$F(\theta) \approx F(\bar{\theta}) + F_\theta(\theta - \bar{\theta}) + \frac{1}{2}(\theta - \bar{\theta})^T F_{\theta\theta}(\theta - \bar{\theta}) \qquad (4.81)$$

where the partial derivatives $F_\theta = \partial F/\partial \theta$ (a row vector) and $[F_{\theta\theta}]_{ij} = \partial^2 F/\partial \theta_i \partial \theta_j$ ($F_{\theta\theta}$ is a $n \times n$ matrix called the Hessian) are computed at $\bar{\theta}$. The values at which $F_\theta = 0$ are called stationary points. From the approximation, we can infer that the necessary conditions for a minimum are:

$$\begin{aligned} F_\theta &= 0 \\ F_{\theta\theta} &\geq 0 \end{aligned} \qquad (4.82)$$

whereas sufficient conditions for a minimum are:

$$\begin{aligned} F_\theta &= 0 \\ F_{\theta\theta} &> 0 \end{aligned} \qquad (4.83)$$

The type of stationary point depends on the Hessian $F_{\theta\theta}$

- If $F_{\theta\theta} > 0$, we have a local minimum.
- If $F_{\theta\theta} < 0$, we have a local maximum.
- If $F_{\theta\theta} \geq$ or $F_{\theta\theta} \leq 0$, we have a singular point.
- If $F_{\theta\theta}$ has both positive and negative eigenvalues, then we have a *saddle point*.

Gradient descent method. It is not difficult to prove that the gradient vector $\nabla F(\theta) = F_\theta^T$ points in the direction of maximum rate of increase of F at θ. Then, the vector $-\nabla F(\theta)$ points in the direction of maximum rate of decrease of F at θ. Let $\theta^{(0)}$ be a starting point, and consider the point $\theta^{(0)} - \alpha \nabla F(\theta^{(0)})$ with $\alpha > 0$. From Taylor's formula, we obtain:

84   4. Dynamic Neural Networks

$$F(\theta^{(0)} - \alpha \nabla F(\theta^{(0)})) \approx F(\theta^{(0)}) - \alpha ||\nabla F(\theta^{(0)})||^2 \tag{4.84}$$

Then for a small α, if $\nabla F(\theta) \neq 0$, we have that $F(\theta^{(0)} - \alpha \nabla F(\theta^{(0)})) < F(\theta^{(0)})$, so that we have achieved an improvement. To formulate an algorithm that implements the above result, suppose that we are given a point $\theta^{(k)}$. To find the next point $\theta^{(k+1)}$, we start at $\theta^{(k)}$ and move by an amount $-\alpha_k \nabla F(\theta^{(k)})$, where α_k is known as the step size. This leads to the following iterative algorithm:

$$\theta^{(k+1)} = \theta^{(k)} - \alpha_k \nabla F(\theta^{(k)}) \tag{4.85}$$

This is known as the *gradient descent algorithm*.

Steepest descent method. This is a gradient algorithm where the step size α_k is chosen to achieve maximum decrease of the objective function at each individual step. At each step α_k is chosen as follows:

$$\alpha_k = \arg \min_{\alpha > 0} F(\theta^{(k)}) - \alpha \nabla F(\theta^{(k)})) \tag{4.86}$$

Line search methods. Most nonlinear programming problems compute at some stage a search direction along which the solution is estimated to lie. The minimum along the line formed by this search direction is generally approximated using a search procedure (e.g. Golden Section), or by polynomial interpolation. Polynomial methods approximate a number of points by a polynomial whose minimum is calculated easily.

Suppose that we know that the minimum of a univariate function $q(\alpha) : \Re \to \Re$ lies in the interval $[a_0, b_0]$. Quadratic interpolation involves a data fit to a univariate function of the form:

$$m(\alpha) = a\alpha^2 + b\alpha + c \tag{4.87}$$

With this approximation, we can estimate that the minimum of $q(\alpha)$ occurs at the minimum of $m(\alpha)$:

$$\alpha^* = \frac{-b}{2a} \tag{4.88}$$

It is possible to calculate the coefficients a, b and c by knowing the values of the function $q(\alpha)$ at three points $\{\alpha_1, \alpha_2, \alpha_3\}$. The value of α that provides the minimum value can be expressed as follows:

$$\alpha^* = \frac{1}{2} \frac{\zeta_{23} q(\alpha_1) + \zeta_{31} q(\alpha_2) + \zeta_{12} q(\alpha_3)}{\eta_{23} q(\alpha_1) + \eta_{31} q(\alpha_2) + \eta_{12} q(\alpha_3)} \tag{4.89}$$

where

$$\zeta_{ij} = \alpha_i^2 - \alpha_j^2 \quad \eta_{ij} = \alpha_i - \alpha_j \tag{4.90}$$

In order to ensure that we calculate a minimum, the following bracketing conditions must be satisfied:

$$q(\alpha_2) < q(\alpha_1) \text{ and } q(\alpha_2) < q(\alpha_3) \tag{4.91}$$

Newton's method. The method of steepest descent uses only first derivatives to search for the minimum of a function. If higher derivatives are used, the resulting iterative algorithm usually performs better. Newton's method uses first and second derivatives. The method is based on a quadratic approximation of the original function that has the same first and second derivatives. It calculates the minimum of the quadratic approximation. It then computes a new quadratic approximation at the new point and computes a new minimum, and so on.

Using a second order Taylor series approximation of F around the current point $\theta^{(k)}$ we obtain:

$$F(\theta) \approx F(\theta^{(k)}) + (\theta - \theta^{(k)})^T g^{(k)} + \tfrac{1}{2}(\theta - \theta^{(k)})^T H^{(k)}(\theta - \theta^{(k)}) = w(\theta) \tag{4.92}$$

where $g^{(k)} = \nabla F(\theta^{(k)})$ and $H^{(k)}$ is the Hessian matrix of F evaluated at $\theta^{(k)}$. To minimise $w(\theta)$, we need:

$$\nabla w(\theta) = g^{(k)} + H(\theta^{(k)})(\theta - \theta^{(k)}) = 0 \tag{4.93}$$

If $H(\theta^{(k)})$ is positive definite, $w(\theta)$ achieves a minimum at:

$$\theta^{(k+1)} = \theta^{(k)} - H(\theta^{(k)})^{-1} g^{(k)} \tag{4.94}$$

Newton's method converges in one step when $F(\theta)$ is a quadratic function. It has better convergence than steepest descent when the starting point is near the solution. However, there are problems when the Hessian matrix is singular. Also, evaluating the Hessian matrix can be computationally expensive. Newton's method does not have the descent property, that is $F(\theta^{(k+1)})$ is not guaranteed to be lower than $F(\theta^{(k)})$, so convergence problems may still occur. We may, however, modify Newton's method to try to guarantee that the algorithm has the descent property by modifying the original method as follows:

$$\theta^{(k+1)} = \theta^{(k)} - \alpha_k H^{-1}(\theta^{(k)})^{-1} g^{(k)} \tag{4.95}$$

where α_k is chosen by means of a line search method to ensure that

$$F(\theta^{(k+1)}) < F(\theta^{(k)}) \tag{4.96}$$

Quasi-Newton algorithms. As mentioned before, one of the disadvantages of Newton's method is the need to calculate the Hessian matrix, as this is a computationally expensive procedure. Quasi-Newton methods are modifications of Newton's method where the Hessian is approximated, so that these algorithms work using only function values and gradients.

In 1970, an inverse Hessian update formula was suggested independently by Broyden, Fletcher, Goldfarb and Shanno. Their formula originated a well known quasi-Newton method. This algorithm can be summarised as follows:

Algorithm 4.6.1 (BFGS Algorithm).

- Step 1: Set $k := 0$, select $\theta^{(0)}$ and a real positive definite matrix B_0.
- If $g^{(k)} = 0$, stop; else set $d^{(k)} = -B_k g^{(k)}$.
- Step 2: Compute

$$\alpha_k = \arg\min_{\alpha > 0} F(\theta^{(k)} + \alpha d^{(k)}) \quad (4.97)$$

$$\theta^{(k+1)} = \theta^{(k)} + \alpha_k d^{(k)} \quad (4.98)$$

- Step 3: Update the inverse Hessian approximation B_{k+1}, set $k := k+1$ and go to step 2.

The inverse Hessian approximation is updated as follows:

$$B_{k+1} = B_k + \left[1 + \frac{\Delta g^{(k)T} B_k \Delta g^{(k)}}{\Delta g^{(k)T} \Delta \theta^{(k)}}\right] \frac{\Delta \theta^{(k)} \Delta \theta^{(k)T}}{\Delta \theta^{(k)T} \Delta g^{(k)}} \\ - \frac{B_k \Delta g^{(k)} \Delta \theta^{(k)T} + (B_k \Delta g^{(k)} \Delta \theta^{(k)T})^T}{\Delta g^{(k)T} \Delta \theta^{(k)}} \quad (4.99)$$

where

$$\Delta \theta^{(k)} = \alpha_k d^{(k)} \quad (4.100)$$

and

$$\Delta g^{(k)} = g^{(k+1)} - g^{(k)} \quad (4.101)$$

Constrained optimisation. A more general class of optimisation problem involves constraints that have to be satisfied at the solution. Equality and inequality constraints are typically considered. A constrained nonlinear optimisation problem is formulated as follows:

$$\min_{\theta} F(\theta)$$

subject to
$$h(\theta) = 0 \quad (4.102)$$
$$g(\theta) \leq 0$$

where $\theta \in \Re^n$ is a decision vector, $F : \Re^n \to \Re$ is an objective function, $h : \Re^n \to \Re^{n_h}$, $g : \Re^n \to \Re^{n_g}$. Several gradient-based methods have been developed to solve this type of problem. A method known as *sequential quadratic programming* (SQP) is perhaps the most widely used technique to solve constrained optimisation problems [104, 105]. Constrained optimisation can be useful for training dynamic neural networks. First, the parameter values can be restricted by means of inequality constraints. Second, relationships between the parameters can be enforced by means of equality constraints in order to fix properties known *a priori*, such as the relative degree of the model (see Example 3.1.4).

4.6.5 Random search methods

Random search methods have the advantage of not requiring the calculation of derivatives of the objective function and are also suitable for training dynamic neural networks. In this subsection, we will introduce an family of optimisation methods inspired by the process of biological evolution called *genetic algorithms*, which are random search methods based on population of solutions and is suitable to locate and deal with the presence of several local minima.

Genetic algorithms. In a broad sense, genetic algorithms proceed as follows. The optimisation procedure starts with a collection of potential solutions – a *population* – each varying somewhat from the other. Each individual of this population is one set of parameters for the parametric model to optimise and they all have a *fitness* or score associated to the performance index (or cost function) to be optimised. In keeping with the analogy of biological evolution the population is evolved by *breeding* the fittest individuals of the population and discarding the unfit ones. During breeding, random alterations – *mutations* – introduce diversity in the solutions. After a number of generations, the fittest individual corresponds to the solution of the optimisation problem.

The stochastic variations introduced by the mutations give the method the potential capability of exploring a vast area of the error surface, which is hard to achieve by using the techniques based on gradient descent. This systematic random search approach is based on evolutionary computation [106] is shown in Figure 4.6. The flowchart provides a general view of the genetic algorithm.

Each individual of the population is a set of parameters, or one potential solution for the parametric model, directly encoded into a unidimensional array called a *chromosome*. The encoding of the parameters into the chromosome must be consistent during the fitting process. More often the chromosome is a binary string but there are cases in which each of the components of the chromosomes or *genes* are real numbers. Hence, each of the genes is one parameter of the model. A chromosome θ represents one set of the L parameters of a given model:

$$\theta = \begin{bmatrix} g_1 & g_2 & \cdots & g_L \end{bmatrix} \tag{4.103}$$

The most commonly used operators for the reproduction of chromosomes into new ones called *offspring* are replication, crossover and mutation.

A *replication* takes place when a chromosome is reproduced without any changes. The resulting chromosome θ_r is identical to its parent θ_f

$$\theta_r = \text{replication}(\theta_f) = \theta_f \tag{4.104}$$

Crossover is a systematic shuffling of genes between two chromosomes. After split points are chosen along the length of one of the chromosomes,

4. Dynamic Neural Networks

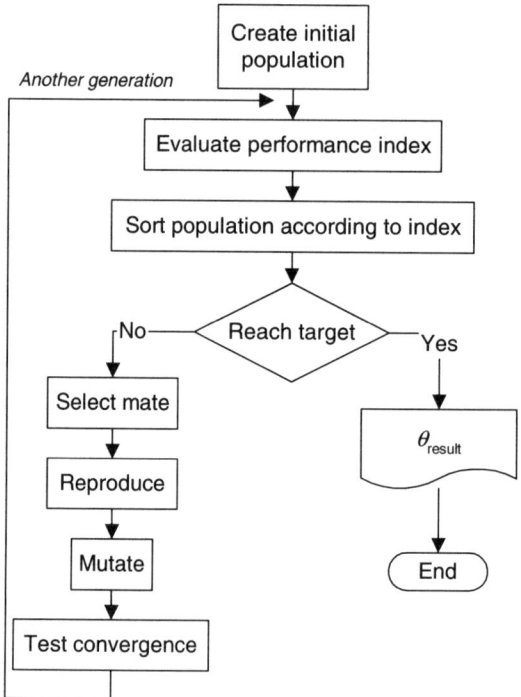

Fig. 4.6. Flowchart of the GA-based training algorithm

the resulting two are formed by swapping the sections separated by the split points. Having the two parents θ_m and θ_f,

$$\theta_m = [p_{m_1}, p_{m_2}, \ldots, p_{m_{i-1}}, p_{m_i}, p_{m_{i+1}}, \ldots, p_{m_{j-1}}, p_{m_j}, p_{m_{j+1}}, \ldots, p_{m_L}]$$
$$\theta_d = [p_{f_1}, p_{f_2}, \ldots, p_{f_{i-1}}, p_{f_i}, p_{f_{i+1}}, \ldots, p_{f_{j-1}}, p_{f_j}, p_{f_{j+1}}, \ldots, p_{f_L}]$$
(4.105)

crossover points i and j are randomly selected, creating three sections in the parents chromosomes $[1, \ldots, i, i+1, \ldots j, j+1, \ldots, L]$. One of the three sections is randomly selected to be combined with its corresponding section from the other parent chromosome. Each offspring parameter value of this section comes from a combination of the two corresponding offspring parameter values,

$$\begin{aligned} p_{a_k} &= \alpha p_{f_k} + (1-\alpha) p_{m_k} \\ p_{b_k} &= (1-\alpha) p_{f_k} + \alpha p_{m_k} \end{aligned}$$
(4.106)

where α is a random number in the interval $[0, 1]$, p_{m_k} is the kth parameter in the mother chromosome and p_{f_k} is the kth parameter in the father chromosome. In the case where the middle section is selected, for example, the resulting offsprings θ_a and θ_b will be defined by,

$$\theta_a = \left[p_{m_1}, p_{m_2}, \ldots, p_{m_{i-1}}, p_{m_i}, p_{a_{i+1}}, \ldots, p_{a_{j-1}}, p_{a_j}, p_{m_{j+1}}, \ldots, p_{m_L}\right]$$
$$\theta_b = \left[p_{f_1}, p_{f_2}, \ldots, p_{f_{i-1}}, p_{f_i}, p_{b_{i+1}}, \ldots, p_{b_{j-1}}, p_{b_j}, p_{f_{j+1}}, \ldots, p_{f_L}\right]$$
(4.107)

The third common operator is called *mutation*. In every generation, some mutations are introduced to keep the genetic algorithm from converging too rapidly to undesirable local minima. With a fixed mutation rate, one element of the offspring chromosomes is randomly changed by a normally distributed random number. A small mutation rate compromises the algorithm's ability to search outside the set of regions covered by the population and rapid convergence to local minima inside such regions.

Algorithm 4.6.2 (Genetic Algorithm).

- Step 1: Create the initial population.
- Step 2: Evaluate the performance index of each individual in the current population.
- Step 3: Sort the population in ascending order according to the value of the performance index of each individual.
- Step 4: Check the stopping criterion (usually a pre-set number of generations). If the stopping criterion is met, then stop, else continue with Step 5.
- Step 5: Randomly select a number of pairs of parents from the current population and carry out crossover, allowing the offpring to replace their parents in the population.
- Step 6: Carry out mutations on a number of randomly selected individuals.
- Step 7: Go back to Step 2.

Genetic algorithm based training of dynamic neural networks. Initially, a population of networks with Gaussian random weights is created. Each individual of the population consists of parameters of a network as given in Equation 4.72 (or as in Equation 4.74), which is illustrated in Figure 4.7. The dimension of the encoded network is $n(n+p+1)$ (or $\{n(n+p+1)+(n-p)\}$ if the initial hidden states are also optimised). A suitable performance index is given by Equation 4.69. The population is subsequently sorted in ascending order according to the value of the performance index for each network.

The first individual is mated with the first 30% of the networks under a crossover operation. In this case, each continuous parameter value, which is randomly initiated in the initial population, is propagated to the next generation only in a different combination. Since each element represents a weight of the DNN, it is essential to add new random values to enrich the selection.

After a given number of generations, the individuals are spread in niches over the objective surface around points that minimise locally the objective

90   4. Dynamic Neural Networks

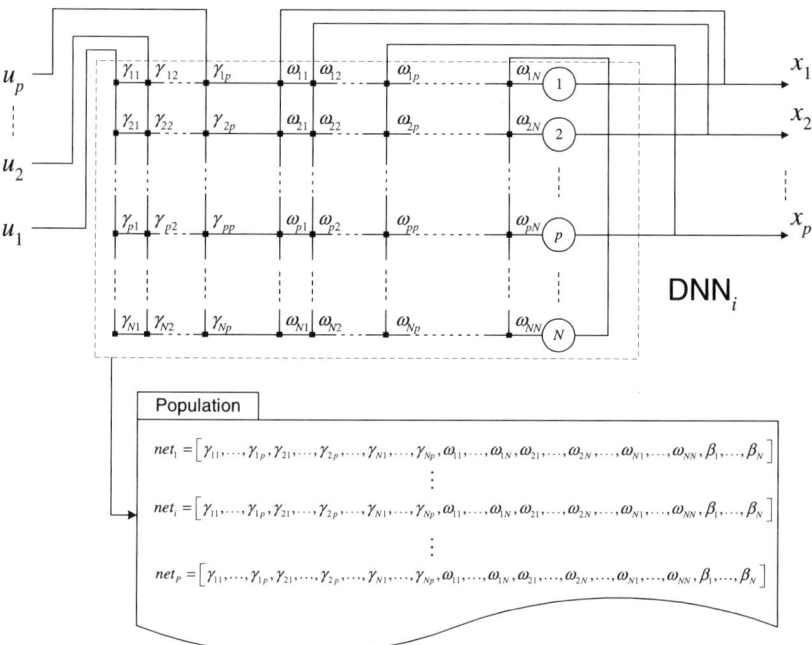

Fig. 4.7. The genetic algorithm training procedure directly encodes each DNN as one individual of the population. The result is decoded back into a DNN configuration

function. The final solution is taken as the fittest individual of the population after a stopping criterion has been met.

Combined algorithms. The use of methods that explore a reduced region of the error surface, such as quasi-Newton methods, typically results in finding a local minimum from given initial parameters. On the other hand, a typical resulting population after several iterations of a genetic algorithm contains individuals that are close to different local minima. A combined algorithm starts with a genetic algorithm based training and then fine-tunes the result using a faster local search procedure [107, 78, 79]. This is illustrated in Figure 4.8.

The genetic algorithm based method allows finding the vicinity of several local minima or niches. Typically, after a few iterations the population starts to cluster around a number of minima. The best individual of each cluster is a potential initial solution for a local algorithm (such as the BFGS algorithm) that will fine-tune the solution. The resulting set of fine-tuned local minima represents a family of DNNs, where each DNN approximates to a different extent the dynamics of the system being modelled.

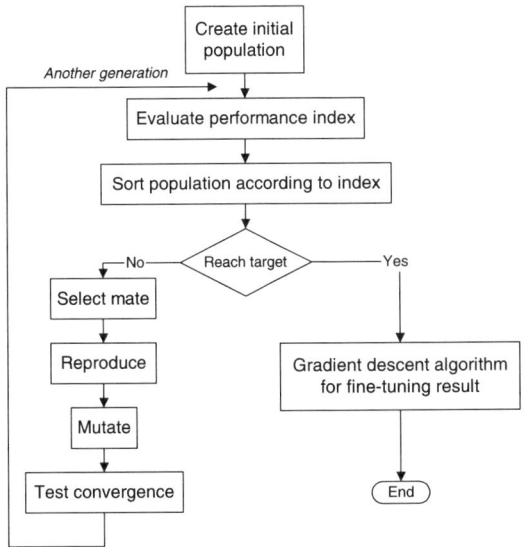

Fig. 4.8. Descriptive diagram of a combined GA–gradient based training algorithm

4.7 Validating the Dynamic Neural Models

Once a model has been trained, it is important to validate it using a data set that should be different from the data used to calculate the model parameters. The part of the data that the model could not reproduce are the residuals:

$$e(t_k) = e(t_k, \theta) = y(t_k) - \hat{y}(t_k, \theta) \qquad (4.108)$$

where $k = 1, \cdots, M$. These residuals carry information about the quality of the model. When doing linear model validation, one typically computes the mean square error (MSE), the autocorrelation of the residuals, and the cross-correlation between the residuals and the input. The values of the autocorrelation and cross-correlation should be small and lie within certain confidence limits. In the nonlinear case, many other correlations may be computed [108].

4.7.1 Overtraining and overfitting

Suppose that we train a dynamic neural network model based on the mean square criterion $F_M(\theta, Z_M)$ and observe the evolution of the training error $F_M(\theta^{(i)}, Z_M)$ and the validation error $F_M(\theta^{(i)}, Z_V)$, where i is an iteration index, Z_M is the training data set and Z_V is the validation data set. A typical plot of these quantities is shown in Figure 4.9. Notice that although the mean square error for the training data reduces monotonically as the iterations progress, the mean square error for the validation data starts to increase at some point during the training [32].

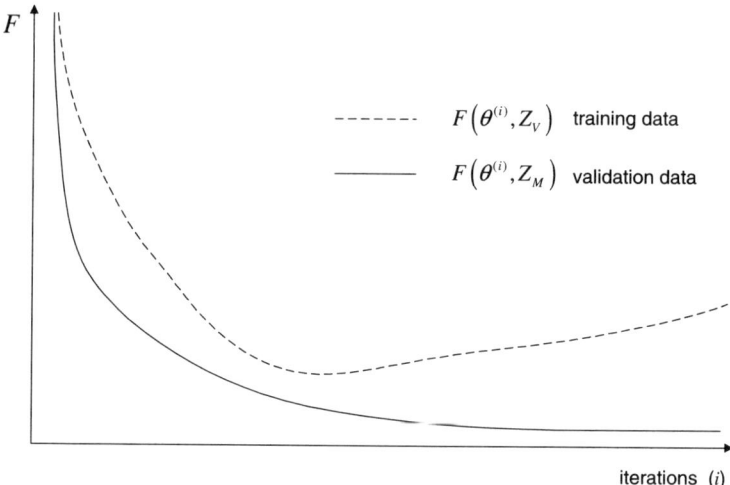

Fig. 4.9. Typical evolution of the mean square error during training

4.7.2 Generalisation, bias and variance

It is interesting to see how good the model is in terms of data not seen by the model during training. To this end, one may define the *generalisation error* as follows [35]:

$$\tilde{F}(\theta, M) = EF_M(\theta, Z_M) = \frac{1}{2}E\left\{||y(t) - \hat{y}(t|\theta)||^2\right\} \quad (4.109)$$

where the expected value is taken with respect to data sets of size M. Now, for $M \to \infty$, we have that the expected objective function becomes:

$$\bar{F}(\theta) = \lim_{M \to \infty} \tilde{F}(\theta, M) \quad (4.110)$$

Let $\hat{\theta}_M$ be the parameter that minimises the objective function for the training data set Z_M:

$$\hat{\theta}_M = \arg\min_\theta F_M(\theta, Z_M) \quad (4.111)$$

Under certain circumstances (see Ljung [31]), it is possible to approximate $\hat{F}$ with a simple formula like Akaike's *final prediction error*:

$$\tilde{F} \approx F_M(\hat{\theta}_M, Z_M) + \lambda_0 \frac{2\dim(\theta)}{M} \quad (4.112)$$

where λ_0 is a constant. Then it is possible to approximate the expected generalisation error for unseen training sets, by adding the second term to the best objective function obtained with the training data set Z_M. This formula can be decomposed in two terms:

$$\tilde{F} \approx F_{\text{bias}} + F_{\text{variance}} \tag{4.113}$$

The bias term is mainly affected by the model structure. A complex and flexible model structure gives a low bias term. The variance term decreases with the amount of data in the training set (M), but it increases with model complexity ($\dim(\theta)$). Then, although increasing the model complexity will reduce the bias error, it will also make the variance error worse. This is known as the bias–variance dilemma.

4.7.3 Cross-validation and model structure selection

Cross-validation is an approach to aid in the model structure selection using an extra data set. Given a training data set Z_M and a validation data set $\mathcal{V}_M$, a sequence of models of increasing complexity $\{\mathcal{M}_1, \mathcal{M}_2, \ldots, \mathcal{M}_k\}$ is identified based on the training data set Z_M. This gives rise to a decreasing trend in the training objective function $F_M(\theta^{(k)}, Z_M)$ for increasing complexity. However, when the objective function is evaluated using the validation data set, $F_M(\theta^{(k)}, \mathcal{V}_M)$, its value will initially decrease with increasing model complexity until some critical k^*, and then it will start to increase. The model $\mathcal{M}_{k^*}$ that provides the minimum value of $F_M(\theta^{(k)}, \mathcal{V}_M)$ is usually chosen as the best model for the system that generated the data.

In the case of the dynamic neural networks that we have studied in this chapter, the model complexity index is the number of neurons in the network; this is, the model order N.

4.7.4 Regularisation

Regularisation techniques allow one to influence the complexity of a model without changing the nominal number of parameters. Regularisation makes a model behave as if it possesses fewer parameters and it really has. Regularisation increases the bias error and decreases the variance error. These techniques are typically applied to neural network models as they tend to have a large number of parameters.

The approach known as *ridge regularisation* modifies the objective function F_M in order to include a term that penalises model complexity [31, 32]. The approach combines training and model structure selection in one optimisation problem. The training criterion is augmented with a penalty term as follows:

$$F_M^R(\theta, Z_M) = F_M(\theta, Z_M) + \frac{1}{2}\rho||\theta||^2 \tag{4.114}$$

where $\rho > 0$ is a penalty factor. The idea behind the additional penalty term $\frac{1}{2}\rho||\theta||^2$ is as follows. Those parameters that are not important in minimising

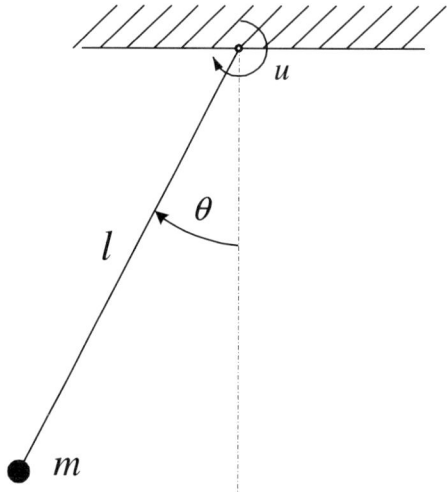

Fig. 4.10. Single link manipulator

the objective function are driven to zero to reduce their contribution to the penalty term, while only the significant parameters will be used since their error reduction effect is larger than their contribution to the penalty term. Other regularisation techniques are discussed in Reference [32].

4.8 A Training Example

Example 4.8.1. Consider the following model for a single link manipulator:

$$\ddot{\theta} = -\frac{g}{l}\sin\theta - \frac{v}{ml^2}\dot{\theta} + \frac{1}{ml^2}u \qquad (4.115)$$

where θ is the angular position, m is the mass of the end of rod element, l is the length of the rod, v is the friction coefficient at the pivot point, and u is the applied torque at the pivot point, see Figure 4.10.

If we define $x_1 = \theta$, $x_2 = \dot{\theta}$, and if $m = 2$ kg, $l = 1$ m and $v = 6$ kg m²/s, the model can be written as follows:

$$\dot{x}_1 = x_2$$
$$\dot{x}_2 = -9.8\sin x_1 - 3x_2 + 0.5u \qquad (4.116)$$
$$y = x_1$$

An identification experiment has been carried out on this single link manipulator model by applying random steps in the input u over a period of time of 100 s. The histories were split into two sets of equal length intended for training and validation. Figure 4.11 shows the input and output trajectories used for training. Figure 4.12 shows the input and output trajectories used for validation. Figure 4.13 shows the results of applying the cross-validation procedure outlined in Section 4.7.3. Training was carried out by means of the BFGS algorithm. Fifty training runs were carried out for each network size, starting from random initial weights, and the best network in terms of the mean square error for the training data was selected for each network size. This was done so as to avoid the effects of local minima and to eliminate the dependency on the initial weights. From Figure 4.13, it is clear that the best performing number of units (or model order) is two, considering the mean square error for the validation data, which coincides with the order of the original system.

The values of the parameters of the best model found were:

$$\beta = \begin{bmatrix} 0.8582 & 0 \\ 0 & 0.7561 \end{bmatrix} \tag{4.117}$$

$$\omega = \begin{bmatrix} 0.6161 & -4.0399 \\ 2.3574 & -2.0194 \end{bmatrix} \tag{4.118}$$

$$\gamma = \begin{bmatrix} -0.0234 \\ -0.1428 \end{bmatrix} \tag{4.119}$$

$$C_n = \begin{bmatrix} 1 & 0 \end{bmatrix} \tag{4.120}$$

Figure 4.14 compares the network output obtained for the training input with the training output. Figure 4.15 compares the network output obtained for the validation input with the validation output. Figure 4.15 shows a further validation input consisting of a square wave, followed by a sinusoidal signal, followed by a sawtooth signal, and compares the network output obtained for that input with the output obtained from the physical model of the single link manipulator. It is clear from these results that the network has learnt well the dynamics of the system.

4.9 Summary

The central aspects of this chapter were the introduction of the dynamic neural network architecture, the study of its properties and the introduction of a number of methods for training dynamic neural networks. The structure of a general artificial neuron was introduced and later we showed how these basic neurons may be configured and combined in different ways to originate various types of neural networks, including static and dynamic forms. The stability of dynamic neural networks was studied for autonomous and non–autonomous networks and sufficient conditions were provided for each case.

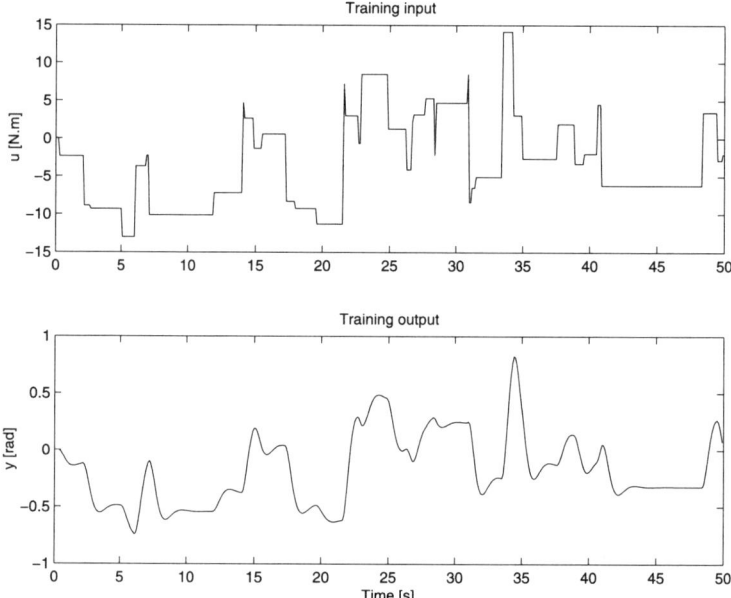

Fig. 4.11. Training data obtained from the single link manipulator model

Also, the problem of training a dynamic neural network was presented as an optimisation problem where the parameters of the network are adjusted by a numerical algorithm that seeks to minimise an error function. General procedures were provided for training and initialising dynamic neural networks. Specific algorithms including various gradient-based methods and a genetic algorithm were discussed for solving the optimisation problems associated with training dynamic neural networks. Furthermore, methods were given for validating the neural networks and selecting their structure.

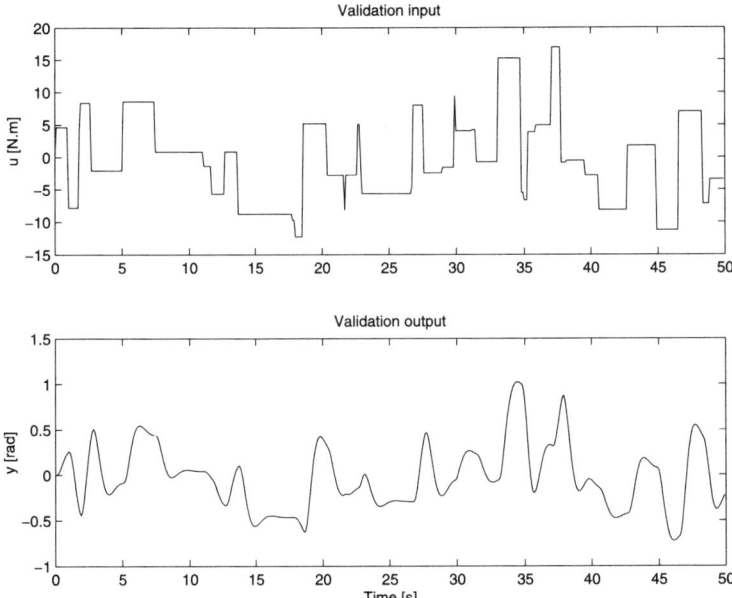

Fig. 4.12. Validation data obtained from the single link manipulator model

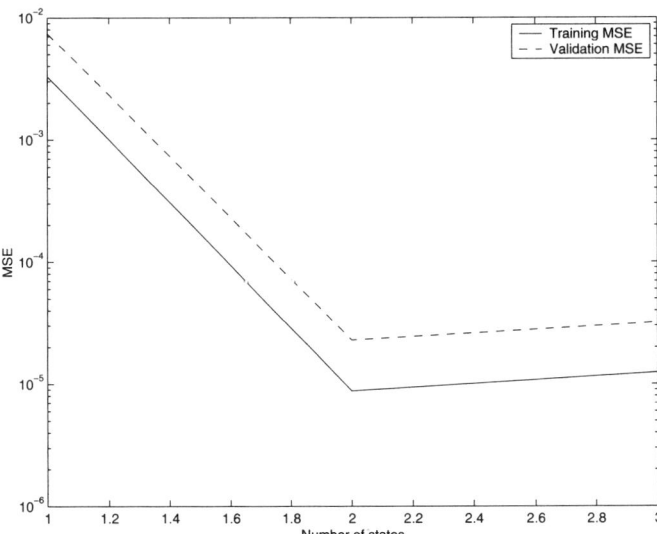

Fig. 4.13. Cross-validation plot for the single link manipulator example

98   4. Dynamic Neural Networks

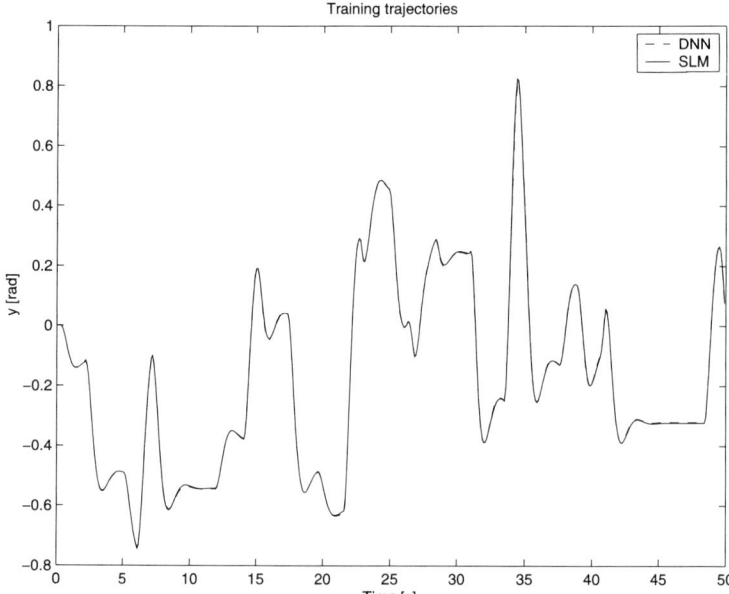

Fig. 4.14. Comparison of training trajectories

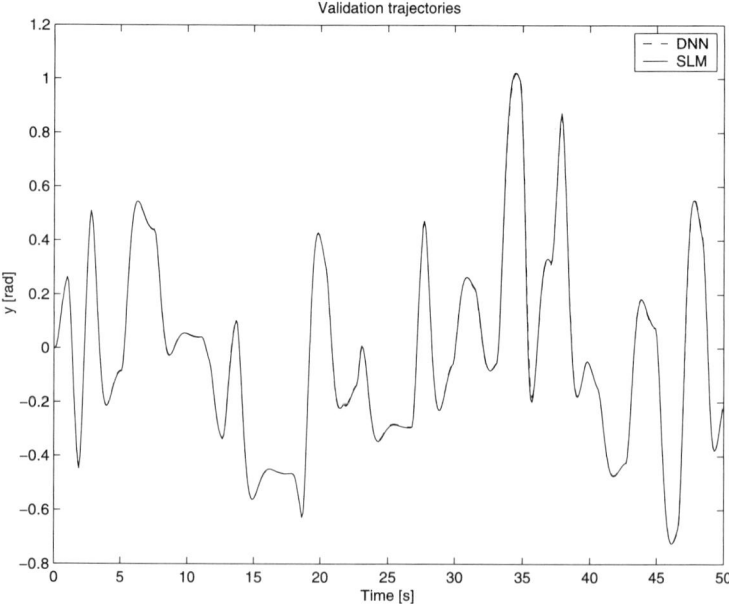

Fig. 4.15. Comparison of validation trajectories

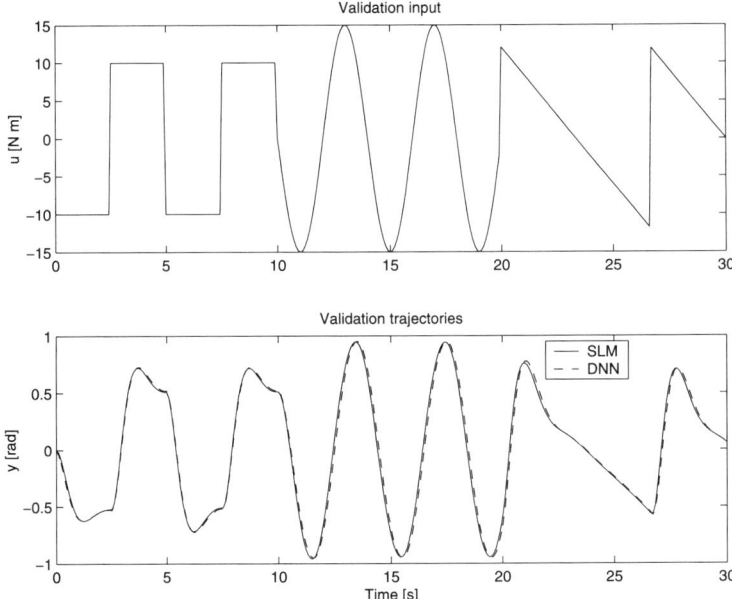

Fig. 4.16. Further validation input and comparison of output trajectories

CHAPTER 5

NONLINEAR SYSTEM APPROXIMATION USING DYNAMIC NEURAL NETWORKS

Identification of dynamic systems is often an important prerequisite for a successful analysis and controller design. Due to the nonlinear nature of most of the processes encountered in engineering applications, there has been extensive research covering the field of nonlinear system identification [109, 110, 111, 112]. It is here that the use of neural networks emerges as a feasible solution. The universal approximation properties of static neural networks [77] make them a useful tool for modelling nonlinear systems. The problem of nonlinear modelling using static neural networks has been extensively researched [33, 113] and many approaches have used multilayer perceptrons and radial basis functions [110, 114, 115].

As discussed in Chapter 4, adding internal dynamics to neural networks and using them for nonlinear system modelling seemed to be a necessary enhancement and several techniques have been proposed that have shown the nonlinear modelling properties of dynamic neural networks [33, 116, 117, 118]. Further work has demonstrated how dynamic neural networks can approximate finite trajectories of n-dimensional autonomous dynamic systems of the form $\dot{x} = f(x)$ [119, 120]. This chapter proves the ability of a class of dynamic neural networks that have different time constants in each neuron to approximate autonomous systems of the form $\dot{x} = f(x)$ and non–autonomous systems of the form $\dot{x} = f(x, u)$. Moreover, an upper bound is found for the norm of the approximation error when modelling general nonlinear systems of the form $\dot{x} = f(x, u)$ by means of dynamic neural networks [79].

5.1 The Universal Approximation Property of Static Multilayer Networks

An important result of approximation theory states that a three-layer feedforward neural network with sigmoidal activation functions in the hidden layer and linear activation functions in the output layer, has the ability to approximate any continuous mapping $f : \Re^n \to \Re^m$ to arbitrary precision, provided the number of units in the hidden layer is sufficiently large.

The following theorem is a version of the fundamental approximation theorem provided by Funahashi [77]. Similar results have been obtained by Cybenko [121] and others.

102   5. Nonlinear System Approximation Using DNNs

Theorem 5.1.1. *Let $\sigma : \Re \to \Re$ be a sigmoid function (a non-constant, increasing and bounded continuous function on $\Re$). Let K be a compact subset of $\Re^n$ and $f : K \to \Re$ be a continuous mapping on K. Then, for arbitrary $\epsilon > 0$, there exists an integer N_h, real constants c_i and b_i, $i = 1, \cdots, N_h$, and w_{ij}, $i = 1, \cdots, N_h$, $j = 1, \cdots, N_h$, such that*

$$\max_{x \in K} \left| f(x) - \sum_{i=1}^{N_h} c_i \sigma \left(\sum_{j=1}^{n} w_{ij} + b_i \right) \right| \leq \epsilon \tag{5.1}$$

The following theorem generalises the above results for vector valued functions $f : \Re^n \to \Re^q$:

Theorem 5.1.2. *Let K be a compact set of $\Re^n$ and $f : K \to \Re^q$ be a continuous mapping. Then, for arbitrary $\epsilon > 0$, there exists an integer N_h, a $q \times N_h$ matrix W_2, an $N_h \times n$ matrix W_1, and an N_h dimensional vector b such that:*

$$\max_{x \in K} \|f(x) - W_2 \sigma(W_1 x + b)\| \leq \epsilon \tag{5.2}$$

where $\sigma : \Re^{N_h} \to \Re^{N_h}$ is a sigmoid mapping whose elements are defined as follows:

$$\sigma(z) = \begin{bmatrix} \sigma(z_1) \\ \vdots \\ \sigma(z_{N_h}) \end{bmatrix} \tag{5.3}$$

where $z = [z_1, \ldots, z_{N_h}]^T \in \Re^{N_h}$.

For the proofs of the above theorems, see [77].

5.2 Dynamic Neural Network Structure

The neural network state space model is defined by a one-dimensional array of neurons; each unit can be described as follows:

$$\dot{x}_i = -\beta_i x_i + \sum_{j=1}^{N} \omega_{ij} \sigma(x_j) + \sum_{j=1}^{p} \gamma_{ij} u_j \tag{5.4}$$

where β_i, ω_{ij} and γ_{ij} are adjustable weights, with $1/\beta_i$ as positive time constant and $p \leq N$, x_i the activation state of the ith unit, $\sigma(\cdot) = \tanh(\cdot)$ and $u_1, \ldots, u_p$ the input signals as seen in Figure 5.1.

The DNN is formed by a single layer of N units. The first p units are taken as the output of the network, leaving $n - p$ units as hidden neurons.

5.3 Approximation Ability of Dynamic Neural Networks

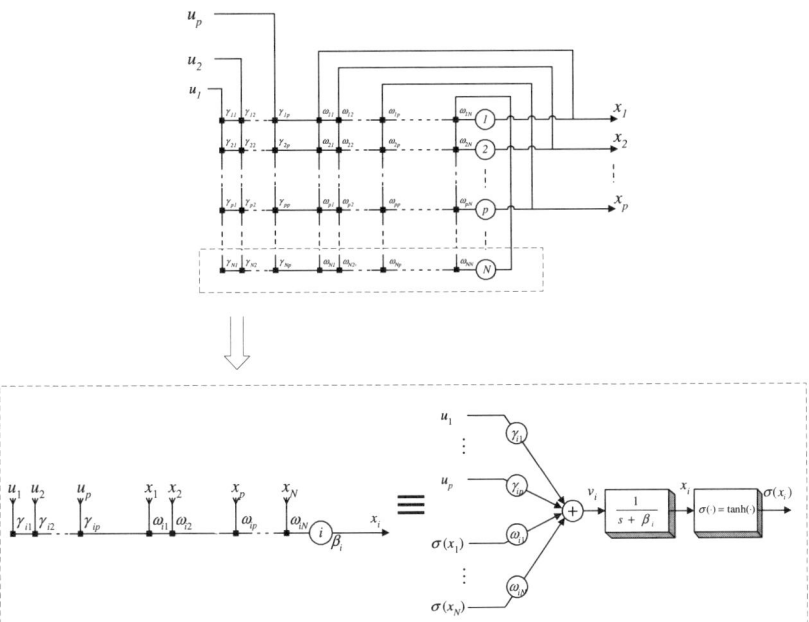

Fig. 5.1. Dynamic neural network with N states, p inputs and p outputs. Descriptive diagram of the dynamic neuron

As shown in Chapter 3, the network is defined by the following vectorised expression:

$$\dot{x} = -\beta x + \omega \sigma(x) + \gamma u$$
$$y_n = C_n x \tag{5.5}$$

5.3 Approximation Ability of Dynamic Neural Networks

As an extension of the problem of approximating time series and trajectories, this section presents an analysis of dynamic system approximation on a Euclidean space using a dynamic neural network. The first section of this chapter shows how a dynamic neural network approximates an autonomous dynamic system. Using the fundamental approximation theorem of neural networks, under certain conditions, the approximation error is bounded to a desired range [119, 120]. Furthermore, the following section investigates the problem of approximating dynamic systems with an external input signal, known as non-autonomous dynamic systems, specifically with systems where the control appears linearly, i.e. control affine systems [107]. Lastly, the closing section uses a Lyapunov approach to extend the identification analysis to general nonlinear systems [79].

5.3.1 Approximation of autonomous nonlinear systems

The following theorem uses the fundamental approximation principle of neural networks [77] to prove that any finite time trajectory of a given n-dimensional autonomous dynamic system can be approximately realised by the internal state of the output units of a continuous time recurrent neural network with n output units, some hidden units and an appropriate initial condition. The analysis is based on the work by Funahashi and Nakamura [119], which has been extended and modified to allow for different time constants $1/\beta_j$ in each neuron.

The following Lemma will be useful for the proof of Theorem 5.3.1. See Chapter 8 of Reference [122] for its proof.

Lemma 5.3.1. *Let $F, \tilde{F} : D \to \Re^n$ be Lipschitz continuous mappings and L be a Lipschitz constant of F. Suppose that for all $x \in D$:*

$$||F(x) - \tilde{F}(x)|| < \varepsilon \qquad (5.6)$$

If $x(t)$ and $\tilde{x}(t)$, are solutions to

$$\begin{aligned} \dot{x} &= F(x) \\ \dot{\tilde{x}} &= \tilde{F}(\tilde{x}) \end{aligned} \qquad (5.7)$$

respectively, on some interval $I = \{t \in \Re | t_0 \le t \le t_f\}$, and $x(t_0) = \tilde{x}(t_0)$, then

$$||x(t) - \tilde{x}(t)|| < \frac{\varepsilon}{L}(\exp(L|t - t_0|) - 1) \qquad (5.8)$$

holds for all $t \in I$.

Theorem 5.3.1. *Let D be an open subset of $\Re^n$, $f : D \to \Re^n$ be a C^1-mapping, and $\tilde{K}$ be a compact subset of D. Suppose that there exists a set $K \subset \tilde{K}$ so that any solution $x(t)$ with initial value $x(0) \in K$ of*

$$\dot{x} = f(x) \qquad (5.9)$$

is defined on $I = [0, T]$ $(0 < T < \infty)$ and is included in $\tilde{K}$ for any $t \in I$. Then, for an arbitrary $\varepsilon > 0$, there is an autonomous dynamic neural network with n output units and N_h hidden units, of the form:

$$\dot{z} = -\beta z + \omega \sigma(z) \qquad (5.10)$$

where $z \in \Re^{n+N_h}$, β is a diagonal matrix and ω is a matrix of the appropriate dimensions, such that for a solution $x(t)$ satisfying Equation 5.9, and an appropriate initial state, the first n units of the network approximate the solution of the autonomous system,

$$\max_{t \in I} ||x(t) - \tilde{x}^o(t)|| < \varepsilon \quad ; \quad I = [0, T] \quad (0 < T < \infty) \qquad (5.11)$$

where $\tilde{x}^o(t) = [\tilde{x}_1^o, \ldots, \tilde{x}_n^o]^T$ is the internal state of output units of the network.

5.3 Approximation Ability of Dynamic Neural Networks

Proof. For given $\varepsilon > 0$, choose η such that $0 < \eta < \min(\varepsilon, \lambda)$, where λ is the distance between $\tilde{K}$ and the boundary ∂D associated with D. Let

$$K_\eta = \left\{ x \in \Re^n; \exists x^* \in \tilde{K}, \|x - x^*\| \leq \eta \right\} \tag{5.12}$$

K_η is a compact subset of D as $\tilde{K}$ is compact. Then f is Lipschitz on K_η. Choose $\epsilon_1 > 0$ so that

$$\epsilon_1 < \frac{\eta l_f}{2(\exp(l_f T) - 1)} \tag{5.13}$$

where l_f is the Lipschitz constant of f on K_η. By the fundamental approximation theorem (Theorem 5.1.2), there exist an integer N_h, matrices $W_2 \in \Re^{n \times N_h}$, $W_1 \in \Re^{N_h \times n}$ and a vector $b \in \Re^{N_h}$, such that

$$\max_{x \in K_\eta} \|f(x) - W_2 \sigma(W_1 x + b)\| \leq \frac{\epsilon_1}{2} \tag{5.14}$$

Define a continuously differentiable mapping $\tilde{f}$ as follows:

$$\tilde{f}(x) = -\tilde{\beta}x + W_2 \sigma(W_1 x + b) \tag{5.15}$$

where $\tilde{\beta} \in \Re^{n \times n}$ is a diagonal matrix with diagonal elements $\{\tilde{\beta}_1, \ldots, \tilde{\beta}_n\}$. Assume that $\tilde{\beta}_j > 0$, $j = 1, \ldots n$, and $\|z\|$, $z \in \Re^{n+N_h}$, are small enough such that:

$$\|\tilde{\beta}x\| < \frac{\epsilon_1}{2} \tag{5.16}$$

$$\|W_1 \tilde{\beta}(W_1^\dagger b - \epsilon_r) - W_3 Q z\| < \frac{\eta l_G}{2(\exp(l_G T) - 1)} \tag{5.17}$$

$$\|\beta\| < \frac{l_G}{2} \tag{5.18}$$

where l_G is a constant to be defined below, β is a $(n + N_h) \times (n + N_h)$ matrix to be defined below, $\epsilon_r \in \Re^n$ is a vector to be defined below, and $W_1^\dagger = (W_1^T W_1)^{-1} W_1^T$ is the Moore–Penrose pseudo-inverse of W_1 (which can also be computed from the singular value decomposition of W_1), W_3 is a $N_h \times N_h$ matrix to be defined below, and Q is a $N_h \times (n + N_h)$ matrix to be defined below. Then, from Equations 5.14 and 5.15:

$$\max_{x \in K_\eta} \|f(x) - \tilde{f}(x)\| \leq \epsilon_1 \tag{5.19}$$

Let $x(t)$ and $\tilde{x}(t)$, $t \in I$, be the solutions of the following equations:

$$\begin{aligned} \dot{x} &= f(x) \\ \dot{\tilde{x}} &= \tilde{f}(\tilde{x}) \end{aligned} \tag{5.20}$$

from the initial condition $x(0) = \tilde{x}(0) = x_0 \in K$, respectively. Thus, by Lemma 5.3.1, for any $t \in I$,

$$\|x(t) - \tilde{x}(t)\| \leq \frac{\epsilon_1}{l_f}(\exp(l_f t) - 1)$$
$$\leq \frac{\epsilon_1}{l_f}(\exp(l_f T) - 1) \tag{5.21}$$

Then, by the inequality in Equation 5.13:

$$\max_{t \in I} \|x(t) - \tilde{x}(t)\| < \frac{\eta}{2} \tag{5.22}$$

Consider now the dynamic system defined by $\tilde{f}$:

$$\dot{\tilde{x}} = -\tilde{\beta}\tilde{x} + W_2 \sigma(W_1 \tilde{x} + b) \tag{5.23}$$

Let $\tilde{y} = W_1 \tilde{x} + b$. Using the Moore-Penrose pseudoinverse of W_1, $\tilde{x}$ can be obtained from $\tilde{y}$ as a least squares solution:

$$\tilde{x} = W_1^\dagger(\tilde{y} - b) + \epsilon_r(\tilde{x}, \tilde{y}) = W_1^\dagger \tilde{y} - W_1^\dagger b + \epsilon_r(\tilde{x}, \tilde{y}) \tag{5.24}$$

where $\epsilon_r \in \Re^n$ is the least squares residual vector. Then

$$\dot{\tilde{y}} = W_1 \dot{\tilde{x}} = -W_1 \tilde{\beta}\tilde{x} + W_1 W_2 \sigma(\tilde{y}) \tag{5.25}$$

using Equation 5.24

$$\dot{\tilde{y}} = -W_1 \tilde{\beta} W_1^\dagger \tilde{y} + W_1 W_2 \sigma(\tilde{y}) + W_1 \tilde{\beta}(W_1^\dagger b - \epsilon_r(\tilde{x}, \tilde{y})) \tag{5.26}$$

Define the following matrices:

$$\bar{\beta} = \text{diag}(W_1 \tilde{\beta} W_1^\dagger) \tag{5.27}$$
$$W_3 = W_1 \tilde{\beta} W_1^\dagger - \bar{\beta} \tag{5.28}$$

so that $\bar{\beta}$ is a diagonal matrix with the diagonal elements of $W_1 \tilde{\beta} W_1^\dagger$, and W_3 is a $N_h \times N_h$ matrix that has zeros on the diagonal and the same elements as $W_1 \tilde{\beta} W_1^\dagger$ off the diagonal. Using $\bar{\beta}$ and W_3, we can rewrite Equation 5.26 as follows:

$$\dot{\tilde{y}} = -\bar{\beta}\tilde{y} + W_1 W_2 \sigma(\tilde{y}) + W_1 \tilde{\beta}(W_1^\dagger b - \epsilon_r(\tilde{x}, \tilde{y})) - W_3 \tilde{y} \tag{5.29}$$

Define a mapping $\tilde{G} : \Re^{n+N_h} \to \Re^{n+N_h}$ as follows:

$$\tilde{G}(\tilde{z}) = -\beta \tilde{z} + \omega \sigma(\tilde{z}) + R(\tilde{z}) \tag{5.30}$$

where

$$\tilde{z} = \begin{bmatrix} \tilde{x} \\ \tilde{y} \end{bmatrix} \in \Re^{n+N_h} \tag{5.31}$$

5.3 Approximation Ability of Dynamic Neural Networks

$$\omega = \begin{bmatrix} 0 & W_2 \\ 0 & W_1 W_2 \end{bmatrix} \in \Re^{(n+N_h) \times (n+N_h)} \tag{5.32}$$

$$\beta = \begin{bmatrix} \tilde{\beta} & 0 \\ 0 & \tilde{\beta} \end{bmatrix} \in \Re^{(n+N_h) \times (n+N_h)} \tag{5.33}$$

$$R(\tilde{z}) = \begin{bmatrix} 0 \\ W_1 \tilde{\beta}(W_1^{\dagger} b - \epsilon_r(\tilde{z})) - W_3 Q \tilde{z} \end{bmatrix} \tag{5.34}$$

where

$$Q = \begin{bmatrix} 0_{N_h \times n} & I_{N_h} \end{bmatrix} \tag{5.35}$$

Define now the mapping $G : \Re^{n+N_h} \to \Re^{n+N_h}$ as follows:

$$G(z) = -\beta z + \omega \sigma(z) \tag{5.36}$$

Then the dynamic system defined by G:

$$\dot{z} = -\beta z + \omega \sigma(z) \tag{5.37}$$

is realised by a dynamic neural network if we set $x^o = [x_1^o, \cdots, x_n^o]^T$ as the internal state of n output units and $x^h = [x_1^h, \cdots, x_{N_h}^h]^T$ as the internal state of N_h hidden units. As G and $\tilde{G}$ are continuously differentiable mappings, and $\partial \sigma(x_j)/\partial x_j$ is bounded for $x_j \in \Re$, then the mapping $z \to \omega\sigma(z)$ is Lipschitz on $\Re^{n+N_h}$ and we can set $l_G/2$ as its Lipschitz constant. Then l_G is the Lipschitz constant of G as the mapping $z \to -\beta z$ has a Lipschitz constant equal to $l_G/2$ by the condition shown in Equation 5.18. Following a similar line of thought, it is easy to find that $\tilde{G}$ is also Lipschitz since the residual $\epsilon_r(\tilde{z})$, and hence $R(\tilde{z})$, are linear functions of $\tilde{z}$. Therefore, Lemma 5.3.1 is applicable to G and $\tilde{G}$.

The difference between $G(z)$ and $\tilde{G}(z)$ for $z \in \Re^{n+N_h}$ is given by:

$$G(z) - \dot{\tilde{G}}(z) = -R(z) \tag{5.38}$$

Taking a norm on both sides and using the condition in Equation 5.17:

$$\|G(z) - \tilde{G}(z)\| = \|R(z)\| = \|W_1 \tilde{\beta}(W_1^{\dagger} b - \epsilon_r(z)) - W_3 Q z\| < \frac{\eta l_G}{2(\exp(l_G T) - 1)} \tag{5.39}$$

Set $\tilde{z}(t)$ and $z(t)$ as the solutions of the following equations, respectively:

$$\dot{\tilde{z}} = \tilde{G}(\tilde{z}); \quad \tilde{x}(0) = x_0 \in K, \quad \tilde{y}(0) = W_1 x_0 + b \tag{5.40}$$

$$\dot{z} = G(z); \quad x^o(0) = x_0 \in K, \quad x^h(0) = W_1 x_0 + b \tag{5.41}$$

Then, by Lemma 5.3.1, and considering Equation 5.39, we have:

$$\max_{t\in I} ||z(t) - \tilde{z}(t)|| \leq \frac{\eta}{2} \tag{5.42}$$

which implies that

$$\max_{t\in I} ||x^o(t) - \tilde{x}(t)|| \leq \frac{\eta}{2} \tag{5.43}$$

where $\tilde{x}(t)$ is the same $\tilde{x}(t)$ as in Equation 5.22. Notice that Equation 5.43 is equivalent to:

$$\max_{t\in I} ||\tilde{x}(t) - x^o(t)|| \leq \frac{\eta}{2} \tag{5.44}$$

Using Equations 5.22 and 5.44, for a given $\varepsilon > 0$, it is possible to construct an autonomous dynamic neural network with internal state $z(t)$, having n output units and N_h hidden units, using the matrices β and w introduced above. For a state trajectory satisfying Equation 5.9, this is:

$$\dot{x} = f(x), \quad x(0) = x_0 \tag{5.45}$$

if the initial state of the network is chosen as follows:

$$x^o(0) = x(0), \quad x^h(0) = W_1 x(0) + b \tag{5.46}$$

then the following is obtained:

$$\begin{aligned}\max_{t\in I} ||x(t) - x^o(t)|| &\leq \max_{t\in I} (||x(t) - \tilde{x}(t)|| + ||\tilde{x}(t) - x^o(t)||) \\ &\leq \max_{t\in I} ||x(t) - \tilde{x}(t)|| + \max_{t\in I} ||\tilde{x}(t) - x^o(t)|| \\ &\leq \tfrac{\eta}{2} + \tfrac{\eta}{2} = \eta\end{aligned} \tag{5.47}$$

But since $\eta < \varepsilon$, then:

$$\max_{t\in I} ||x(t) - x^o(t)|| < \varepsilon \tag{5.48}$$

which completes the proof. □.

Remark 5.3.1. Notice the importance of the hidden units in the approximation ability of the dynamic neural network.

Remark 5.3.2. If all diagonal elements of β are equal, then the theorem proved in [119] applies.

Remark 5.3.3. If $n = N_h$ and W_1 has full rank, then Theorem 5.3.1 applies with no least squares residual ($\epsilon_r(z) = 0$), even if the diagonal elements of β are not equal.

Remark 5.3.4. Notice that the dynamic neural network constructed as part of the proof of Theorem 5.3.1 has a matrix w with the special structure given by Equation 5.32.

Remark 5.3.5. In the proof by Funahashi [119], there was no need to use a pseudo-inverse of W_1 to obtain $\dot{y}$ in terms of y (see Equation 5.26), since he was dealing with a special case where β is a diagonal matrix with equal elements on the diagonal ($\tilde{\beta} = \frac{1}{\tau}I$). This corresponds with the case where the time constants associated with neurons are equal. In the case treated in Theorem 5.3.1, the time constants of neurons are allowed to be different.

5.3.2 Approximation of non-autonomous nonlinear systems

Based on the approximation of autonomous nonlinear systems [119, 120], this section describes how any finite time trajectory of a given finite-dimensional non–autonomous dynamic system $\dot{x}(t) = f(x(t), u(t))$ can be approximated by the internal state of the output units of a continuous time dynamic neural network.

Corollary 5.3.1. *Let K and U be compact subsets of $\Re^n$ and $\Re^m$, respectively, and $f : K \times U \to \Re^n$ be a continuous mapping. Then, for arbitrary $\epsilon > 0$, there exists an integer N_h, an $n \times N_h$ matrix W_2, an $N_h \times n$ matrix W_1, an $N_h \times m$ matrix γ_1, and an N_h dimensional vector b such that:*

$$\max_{x \in K, u \in U} \|f(x, u) - W_2 \sigma(W_1 x + \gamma_1 u + b)\| \le \epsilon \tag{5.49}$$

where $\sigma : \Re^{N_h} \to \Re^{N_h}$ is a sigmoid mapping whose elements are defined as follows:

$$\sigma(z) = \begin{bmatrix} \sigma(z_1) \\ \vdots \\ \sigma(z_{N_h}) \end{bmatrix} \tag{5.50}$$

where $z = [z_1, \ldots, z_{N_h}]^T \in \Re^{N_h}$.

Proof. The proof follows directly from Theorem 5.1.2, by making the following substitutions: $S \leftarrow S \times U$, $q \leftarrow n$, $x \leftarrow [x, u]^T$, $W_1 \leftarrow [W_1 \; \gamma_1]$.

The following lemma will be useful. Its proof is similar to that of Lemma 5.3.1, provided in [122].

Lemma 5.3.2. *Let $F, \tilde{F} : S \times U \to \Re^n$ be Lipschitz continuous mappings and L be a Lipschitz constant of $F(x, u)$ in x on $S \times U$. Suppose that for all $x \in S$ and $u \in U$:*

$$\|F(x, u) - \tilde{F}(x, u)\| < \varepsilon \tag{5.51}$$

If $x(t)$ and $\tilde{x}(t)$, are solutions to

$$\begin{aligned} \dot{x} &= F(x, u) \\ \dot{\tilde{x}} &= \tilde{F}(\tilde{x}, u) \end{aligned} \tag{5.52}$$

respectively, on some interval $I = \{t \in \Re | t_0 \leq t \leq t_f\}$, and $x(t_0) = \tilde{x}(t_0)$, then

$$||x(t) - \tilde{x}(t)|| < \frac{\varepsilon}{L}\left(\exp(L|t - t_0|) - 1\right) \quad (5.53)$$

holds for all $t \in I$.

Theorem 5.3.2. *Let D be an open subset of $\Re^n$, and U and open subset of $\Re^m$. Let $f : D \times U \to \Re^n$ be a C^1-mapping, $u : [0, T] \to U$ be a C^1 function, $\tilde{K}$ be a compact subset of D and $\tilde{U}$ be a compact subset of U. Suppose that there exists a set $K \subset \tilde{K}$ so that any solution $x(t)$ with initial value $x(0) \in K$ of*

$$\dot{x}(t) = f(x(t), u(t)) \quad (5.54)$$

is defined on $I = [0, T]$ $(0 < T < \infty)$ for $u(t) \in U$ with $t \in I$, and is included in $\tilde{K}$ for any $t \in I$. Then, for an arbitrary $\varepsilon > 0$, there is a non-autonomous dynamic neural network with n output units and N_h hidden units, of the form:

$$\dot{z} = -\beta z + \omega \sigma(z) + \gamma \bar{u} \quad (5.55)$$

where $z \in \Re^{n+N_h}$, $\bar{u} = [u \; \dot{u}]^T \in \Re^{2m}$, β is a diagonal matrix, ω and γ are matrices of the appropriate dimensions, such that for a solution $x(t)$ satisfying Equation 5.54, and an appropriate initial state, the first n units of the network approximate the solution of the autonomous system,

$$\max_{t \in I} ||x(t) - \tilde{x}^o(t)|| < \varepsilon \quad ; \quad I = [0, T] \quad (0 < T < \infty) \quad (5.56)$$

where $\tilde{x}^o(t) = [\tilde{x}_1^o, \ldots, \tilde{x}_n^o]^T$ is the internal state of output units of the network.

Proof. For given $\varepsilon > 0$, choose η such that $0 < \eta < \min(\varepsilon, \lambda_1, \lambda_2)$, where λ_1 is the distance between $\tilde{K}$ and the boundary ∂D of D and λ_2 is the distance between $\tilde{U}$ and the boundary ∂U of U. Let

$$K_\eta = \left\{x \in \Re^n; \exists x^* \in \tilde{K}, ||x - x^*|| \leq \eta\right\} \quad (5.57)$$

and

$$U_\eta = \left\{u \in \Re^m; \exists u^* \in \tilde{U}, ||u - u^*|| \leq \eta\right\} \quad (5.58)$$

K_η and U_η are compact subsets of D and U, respectively, as $\tilde{K}$ and $\tilde{U}$ are compact. Then f is Lipschitz on $K_\eta \times U_\eta$ (see [123]). Choose $\epsilon_1 > 0$ so that

$$\epsilon_1 < \frac{\eta l_f}{2(\exp(l_f T) - 1)} \quad (5.59)$$

5.3 Approximation Ability of Dynamic Neural Networks

where l_f is the Lipschitz constant of f on K_η. By Corollary 5.3.1, there exist an integer N_h, matrices $W_2 \in \Re^{n \times N_h}$, $W_1 \in \Re^{N_h \times n}$ and a vector $b \in \Re^{N_h}$, such that

$$\max_{x \in K_\eta} \|f(x,u) - W_2\sigma(W_1 x + \gamma_1 u + b)\| \leq \frac{\epsilon_1}{2} \tag{5.60}$$

Define a continuously differentiable mapping $\tilde{f}$ as follows:

$$\tilde{f}(x,u) = -\tilde{\beta}x + \gamma_2 u + W_2 \sigma(W_1 x + \gamma_1 u + b) \tag{5.61}$$

where $\tilde{\beta} \in \Re^{n \times n}$ is a diagonal matrix with diagonal elements $\{\tilde{\beta}_1, \ldots, \tilde{\beta}_n\}$ and $\gamma_2 \in \Re^{n \times m}$. Assume that $\tilde{\beta}_j > 0$, $j = 1, \ldots n$, $\|u\|$ and $\|z\|$, $z \in \Re^{n+N_h}$, are small enough such that:

$$\|\tilde{\beta}x - \gamma_2 u\| < \frac{\epsilon_1}{2} \tag{5.62}$$

$$\|W_1 \tilde{\beta}(W_1^\dagger b - \epsilon_r) - W_3 Q z\| < \frac{\eta l_G}{2(\exp(l_G T) - 1)} \tag{5.63}$$

$$\|\beta\| < \frac{l_G}{2} \tag{5.64}$$

where l_G is a constant to be defined below, β is a $(n + N_h) \times (n + N_h)$ matrix to be defined below, $\epsilon_r \in \Re^n$ is a vector to be defined below, and $W_1^\dagger = (W_1^T W_1)^{-1} W_1^T$ is the Moore–Penrose pseudo-inverse of W_1 (which can also be computed from the singular value decomposition of W_1), W_3 is a $N_h \times N_h$ matrix to be defined below, and Q is a $N_h \times (n + N_h)$ matrix to be defined below. Then, from Equations 5.60 and 5.61:

$$\max_{x \in K_\eta, u \in U_\eta} \|f(x,u) - \tilde{f}(x,u)\| \leq \epsilon_1 \tag{5.65}$$

Let $x(t)$ and $\tilde{x}(t)$, $t \in I$, be the solutions of the following equations:

$$\dot{x}(t) = f(x(t), u(t))$$
$$\dot{\tilde{x}}(t) = \tilde{f}(\tilde{x}(t), u(t)) \tag{5.66}$$

from the initial condition $x(0) = \tilde{x}(0) = x_0 \in K$, respectively. Thus, by Lemma 5.3.2, for any $t \in I$,

$$\|x(t) - \tilde{x}(t)\| \leq \tfrac{\epsilon_1}{l_f}(\exp(l_f t) - 1)$$
$$\leq \tfrac{\epsilon_1}{l_f}(\exp(l_f T) - 1) \tag{5.67}$$

Then, by the inequality in Equation 5.59:

112   5. Nonlinear System Approximation Using DNNs

$$\max_{t \in I} ||x(t) - \tilde{x}(t)|| < \frac{\eta}{2} \tag{5.68}$$

Consider now the dynamic system defined by $\tilde{f}$:

$$\dot{\tilde{x}} = -\tilde{\beta}\tilde{x} + W_2\sigma(W_1\tilde{x} + \gamma_1 u + b) + \gamma_2 u \tag{5.69}$$

Let $\tilde{y} = W_1\tilde{x} + \gamma_1 u + b$. Using the Moore–Penrose pseudo-inverse of W_1, $\tilde{x}$ can be obtained from $\tilde{y}$ as a least squares solution:

$$\tilde{x} = W_1^\dagger(\tilde{y} - b - \gamma_1 u) + \epsilon_r(\tilde{x}, \tilde{y}, u) = W_1^\dagger \tilde{y} - W_1^\dagger \gamma_1 u - W_1^\dagger b + \epsilon_r(\tilde{x}, \tilde{y}, u) \tag{5.70}$$

where $\epsilon_r \in \Re^n$ is the least squares residual vector. Then

$$\dot{\tilde{y}} = W_1\dot{\tilde{x}} + \gamma_1 \dot{u} = -W_1\tilde{\beta}\tilde{x} + W_1 W_2 \sigma(\tilde{y}) + W_1\gamma_2 u + \gamma_1 \dot{u} \tag{5.71}$$

using Equation 5.70

$$\dot{\tilde{y}} = -W_1\tilde{\beta}W_1^\dagger \tilde{y} + W_1 W_2 \sigma(\tilde{y}) + W_1\gamma_2 u + W_1\tilde{\beta}(W_1^\dagger b + W_1^\dagger \gamma_1 u - \epsilon_r) + \gamma_1 \dot{u} \tag{5.72}$$

Define the following matrices:

$$\bar{\beta} = \text{diag}(W_1\tilde{\beta}W_1^\dagger) \tag{5.73}$$
$$W_3 = W_1\tilde{\beta}W_1^\dagger - \bar{\beta} \tag{5.74}$$

so that $\bar{\beta}$ is a diagonal matrix with the diagonal elements of $W_1\tilde{\beta}W_1^\dagger$, and W_3 is a $N_h \times N_h$ matrix that has zeros on the diagonal and the same elements as $W_1\tilde{\beta}W_1^\dagger$ off the diagonal. Using $\bar{\beta}$ and W_3, we can rewrite Equation 5.72 as follows:

$$\dot{\tilde{y}} = -\bar{\beta}\tilde{y} + W_1 W_2 \sigma(\tilde{y}) + W_1\gamma_2 u + W_1\tilde{\beta}(W_1^\dagger b + W_1^\dagger \gamma_1 u - \epsilon_r) - W_3\tilde{y} + \gamma_1 \dot{u} \tag{5.75}$$

Define a mapping $\tilde{G} : \Re^{n+N_h} \times \Re^{2m} \to \Re^{n+N_h}$ as follows:

$$\tilde{G}(\tilde{z}, \bar{u}) = -\beta\tilde{z} + \omega\sigma(\tilde{z}) + \gamma\bar{u} + R(\tilde{z}, \bar{u}) \tag{5.76}$$

where

$$\tilde{z} = \begin{bmatrix} \tilde{x} \\ \tilde{y} \end{bmatrix} \in \Re^{n+N_h} \tag{5.77}$$

$$\bar{u} = \begin{bmatrix} u \\ \dot{u} \end{bmatrix} \in \Re^{2m} \tag{5.78}$$

$$\omega = \begin{bmatrix} 0 & W_2 \\ 0 & W_1 W_2 \end{bmatrix} \in \Re^{(n+N_h) \times (n+N_h)} \tag{5.79}$$

$$\beta = \begin{bmatrix} \tilde{\beta} & 0 \\ 0 & \bar{\beta} \end{bmatrix} \in \Re^{(n+N_h) \times (n+N_h)} \tag{5.80}$$

5.3 Approximation Ability of Dynamic Neural Networks

$$\gamma = \begin{bmatrix} \gamma_2 & 0 \\ (W_1\tilde{\beta}W_1^\dagger\gamma_1 + W_1\gamma_2) & \gamma_1 \end{bmatrix} \in \Re^{(n+N_h)\times 2m} \tag{5.81}$$

$$R(\tilde{z}, \bar{u}) = \begin{bmatrix} 0 \\ W_1\tilde{\beta}(W_1^\dagger b - \epsilon_r) - W_3 Q\tilde{z} \end{bmatrix} \in \Re^{n+N_h} \tag{5.82}$$

where

$$Q = \begin{bmatrix} 0_{N_h \times n} & I_{N_h} \end{bmatrix} \tag{5.83}$$

Define now the mapping $G : \Re^{n+N_h} \times \Re^{2m} \to \Re^{n+N_h}$ as follows:

$$G(z, \bar{u}) = -\beta z + \omega\sigma(z) + \gamma\bar{u} \tag{5.84}$$

Then the dynamic system defined by G:

$$\dot{z} = -\beta z + \omega\sigma(z) + \gamma\bar{u} \tag{5.85}$$

is realised by a dynamic neural network if we set $x^o = [x_1^o, \cdots, x_n^o]^T$ as the internal state of n output units and $x^h = [x_1^h, \cdots, x_{N_h}^h]^T$ as the internal state of N_h hidden units. As G and $\tilde{G}$ are continuously differentiable mappings, and $\partial\sigma(x_j)/\partial x_j$ is bounded for $x_j \in \Re$, then the mapping $z \to \omega\sigma(z)$ is Lipschitz on $\Re^{n+N_h}$ and we can set $l_G/2$ as its Lipschitz constant. Then l_G is the Lipschitz constant of G in z as the mapping $z \to -\beta z$ has a Lipschitz constant equal to $l_G/2$ by the condition in Equation 5.64. It is not difficult to infer that $\tilde{G}$ is also Lipschitz, so that Lemma 5.3.2 is applicable to G and $\tilde{G}$.

The difference between $G(z, \bar{u})$ and $\tilde{G}(z, \bar{u})$ for $z \in \Re^{n+N_h}$ is given by:

$$G(z, \bar{u}) - \tilde{G}(z, \bar{u}) = -R(z, \bar{u}) \tag{5.86}$$

Taking a norm on both sides and using the condition in Equation 5.63:

$$||G(z, \bar{u}) - \tilde{G}(z, \bar{u})|| = ||R(z, \bar{u})|| = ||W_1\tilde{\beta}(W_1^\dagger b - \epsilon_r) - W_3 Q z|| \tag{5.87}$$
$$< \frac{\eta l_G}{2(\exp(l_G T) - 1)}$$

Set $\tilde{z}(t)$ and $z(t)$ as the solutions of the following equations, respectively:

$$\dot{\tilde{z}} = \tilde{G}(\tilde{z}, \bar{u}); \quad \tilde{x}(0) = x_0 \in K, \quad \tilde{y}(0) = W_1 x_0 + \gamma_1 u(0) + b \tag{5.88}$$

$$\dot{z} = G(z, \bar{u}); \quad x^o(0) = x_0 \in K, \quad x^h(0) = W_1 x_0 + \gamma_1 u(0) + b \tag{5.89}$$

Then, by Lemma 5.3.2, and considering Equation 5.87, we have:

$$\max_{t \in I} ||z(t) - \tilde{z}(t)|| \leq \frac{\eta}{2} \tag{5.90}$$

which implies that

$$\max_{t\in I} ||x^o(t) - \tilde{x}(t)|| \leq \frac{\eta}{2} \tag{5.91}$$

where $\tilde{x}(t)$ is the same $\tilde{x}(t)$ as in Equation 5.68. Notice that Equation 5.91 is equivalent to:

$$\max_{t\in I} ||\tilde{x}(t) - x^o(t)|| \leq \frac{\eta}{2} \tag{5.92}$$

Using Equations 5.68 and 5.92, for a given $\varepsilon > 0$, it is possible to construct a non-autonomous dynamic neural network with internal state $z(t)$, input $\bar{u}(t)$, having n output units and N_h hidden units, using the matrices β, ω and γ introduced above. For a state trajectory satisfying Equation 5.54, this is:

$$\dot{x}(t) = f(x(t), u(t)), \quad x(0) = x_0 \tag{5.93}$$

if the initial state of the network is chosen as follows:

$$x^o(0) = x(0), \quad x^h(0) = W_1 x(0) + \gamma_1 u(0) + b \tag{5.94}$$

then the following is obtained:

$$\begin{aligned}\max_{t\in I} ||x(t) - x^o(t)|| &\leq \max_{t\in I} (||x(t) - \tilde{x}(t)|| + ||\tilde{x}(t) - x^o(t)||) \\ &\leq \max_{t\in I} ||x(t) - \tilde{x}(t)|| + \max_{t\in I} ||\tilde{x}(t) - x^o(t)|| \\ &\leq \tfrac{\eta}{2} + \tfrac{\eta}{2} = \eta\end{aligned} \tag{5.95}$$

But since $\eta < \varepsilon$, then:

$$\max_{t\in I} ||x(t) - x^o(t)|| < \varepsilon \tag{5.96}$$

which completes the proof. □

Remark 5.3.6. A similar theorem for non-autonomous systems but that employs a different network architecture was proved by Chow and Li [123].

Remark 5.3.7. Notice the special input structure of the network that resulted from this constructive proof and the requirement that $u(t)$ be differentiable with respect to time. In practice, it is usually not necessary to provide input derivative information to achieve a good approximation of the system.

5.3.3 Upper bound on the approximation error of general nonlinear systems

The approach described in this section is based on the analysis presented in [124], where the approximation error is found for nonaffine networks with the same state space dimension as the system [87]. Several modifications are carried out in order to apply their analysis to larger dimension control affine

5.3 Approximation Ability of Dynamic Neural Networks

networks in Equation 5.5, i.e. $N \geqslant n$. Convergence conditions are given by means of a method similar to Lyapunov's second method.

Consider the dynamic neural network described by:

$$\dot{x}^* = -\beta x^* + \omega \sigma(x^*) + \gamma u \qquad (5.97)$$

conformed by n outputs and h internal hidden states as follows:

$$\dot{x}^* = \begin{bmatrix} \dot{x}^o \\ \dot{x}^h \end{bmatrix} = \begin{bmatrix} \beta^o & 0 \\ 0 & \beta^h \end{bmatrix} \begin{bmatrix} x^o \\ x^h \end{bmatrix} + \begin{bmatrix} \omega_a & \omega_b \\ \omega_c & \omega_d \end{bmatrix} \begin{bmatrix} \sigma(x^o) \\ \sigma(x^h) \end{bmatrix} + \begin{bmatrix} \gamma^o \\ \gamma^h \end{bmatrix} u \qquad (5.98)$$

where $x^o \in \Re^n$ and $x^h \in \Re^{N-n}$. On the other hand, consider a general nonlinear system described by

$$\dot{x} = f(x, u) \qquad (5.99)$$

where $x \in \Re^n$ is the system state, $u \in \Re^p$ the external input and $f(\cdot, \cdot)$ a smooth vector field. Suppose that the dynamic neural network in Equation 5.97 has been trained so that its n output states approximate the n states of the system given by Equation 5.99. The approximation error and its time derivative are given by:

$$e = x - x^o$$
$$\dot{e} = \dot{x} - \dot{x}^o \qquad (5.100)$$

Substituting the corresponding expressions for $\dot{x}$ and $\dot{x}^o$ and adding and subtracting Ae, where A is any matrix having all eigenvalues with negative real part,

$$\dot{e} = Ae + \left[f(x, u) - \beta^o x - \omega_a \sigma(x^o) - \omega_b \sigma(x^h) - \gamma^o u - Ae \right] \qquad (5.101)$$

The rate of change in the approximation error is given by,

$$\dot{e} = Ae + h(e, x, x^h, u) \qquad (5.102)$$

where

$$h(e, x, x^h, u) = f(x, u) - \beta^o x - (A - \beta^o)e - \omega_a \sigma(x - e) - \omega_b \sigma(x^h) - \gamma^o u \qquad (5.103)$$

The following assumptions about the system given by Equation 5.99 and the network described by Equation 5.97 are stated:

Assumption 5.3.1. There exist positive definite matrices H_n and H_e [125], such that,

$$h^T(e, x, x^h, u) H_n h(e, x, x^h, u) \leqslant \varepsilon_0(x, u) + \varepsilon_1(x, x^h) e^T H_e e \qquad (5.104)$$

for $\varepsilon_0(\cdot, \cdot)$ and $\varepsilon_1(\cdot, \cdot)$ bounded positive functions that satisfy:

$$\sup_{x,u} \|\varepsilon_0(x, u)\| < \varepsilon^0, \quad \sup_{x, x^h} \|\varepsilon_1(x, x^h)\| < \varepsilon^1 \qquad (5.105)$$

for $x \in \Re^n$, $u \in \Re^p$, $x^h \in \Re^{N-p}$ and $t \geq 0$.

Assumption 5.3.2. The vector field $f(\cdot,\cdot)$ satisfies

$$\|f(x,u) - \beta^\circ x - w_a \sigma(x) - \gamma^\circ u\| \leqslant l_f \quad \text{for all } x \in \Re^n, \text{ and } u \in \Re^p \quad (5.106)$$

Assumption 5.3.3. The vector activation function $\sigma(x)$ is Lipschitz with constant l_σ.

Assumption 5.3.4. The external input u is bounded

$$\|u\| \leqslant \bar{u} \quad (5.107)$$

Assumption 5.3.5. There exists a strictly positive definite matrix Q_0 such that the following matrix Riccati equation,

$$\begin{aligned} A^T P + PA + PRP + Q &= 0 \\ R \equiv H_n^{-1}, \; Q \equiv \varepsilon^1 H_e + Q_0 \end{aligned} \quad (5.108)$$

has a solution $P = P^T > 0$. This condition is fulfilled by selecting A diagonal.

If $h(\cdot,\cdot,\cdot,\cdot)$ is separated into,

$$h(e, x, x^h, u) = h_1(X, u) + h_2(e, X, x^h) \quad (5.109)$$

such that

$$\begin{aligned} h_1(x, u) &= f(x, u) - \beta^\circ x - w_a \sigma(x) - \gamma^\circ u \\ h_2(x, x^h, e) &= w_a[\sigma(x) - \sigma(x - e)] - w_b \sigma(x^h) - (A - \beta^\circ)e \end{aligned} \quad (5.110)$$

then

$$h^T H_n h = \|h\|_{H_n}^2 = \|h_1 + h_2\|_{H_n}^2 \leqslant (1+\chi)\|h_1\|_{H_n}^2 + (1+\chi^{-1})\|h_2\|_{H_n}^2 \quad (5.111)$$

Using Equation 5.111, for any fixed $\chi > 0$ and any positive definite matrix H_n, assuming $H_e = I_n$ (the identity matrix of size $n \times n$), and employing also Assumptions 5.3.2 and 5.3.3, it is possible to find suitable functions $\varepsilon_0(x, u)$ and $\varepsilon_1(x, x^h)$ that satisfy Equation 5.104, and also a suitable value for the bound ε^0:

$$\begin{aligned} \varepsilon_0(x, u) &= (1+\chi)\|f(x,u) - \beta_n x - w_a \sigma(x) - \gamma^\circ u\|_{H_n}^2 \\ \varepsilon_1(x, x^h) &= \left(1+\chi^{-1}\right)\left(l_\sigma\|w_a\| + l_\sigma\|w_b\| + \|A - \beta^\circ\|\right)^2 \|H_n\| \\ \varepsilon^0 &= (1+\chi)\|H_n\| l_f^2 \end{aligned} \quad (5.112)$$

where l_σ is the Lipschitz constant of the vector activation function $\sigma(x)$.

5.3 Approximation Ability of Dynamic Neural Networks

Theorem 5.3.3. *Under Assumptions 5.3.1 to 5.3.5 and given Equation 5.112, the dynamic neural network described by Equation 5.97 approximates the general nonlinear system described by Equation 5.99, with the approximation error*

$$e(t) = x(t) - x^o(t) \tag{5.113}$$

bounded by,

$$\sup \|e(t)\| \leqslant \sqrt{\frac{\eta}{\lambda_{\min}(P)}} \tag{5.114}$$

with

$$R_P = P^{-1/2} Q_o P^{-1/2} \tag{5.115}$$

$$\eta = \max\left(V(e(0)), \frac{\varepsilon^0}{\lambda_{\min}(R_p)}\right) \tag{5.116}$$

$$V(e(t)) = e(t)^T P e(t) \tag{5.117}$$

Proof. Consider the positive definite function $V : \Re^n \to \Re^+$ such that,

$$V(e) = e^T P c, \ P = P^T > 0, \ P \in \Re^{n \times n} \tag{5.118}$$

Calculating its time derivative along the solutions of Equation 5.101 and using the function h given in Equation 5.103, the following is obtained,

$$\dot{V}(e) = 2e^T P(Ae + h) = e^T(PA + A^T P)e + 2e^T Ph \tag{5.119}$$

Using the matrix inequality,

$$Z^T Y + Y^T \leqslant Z^T L Z + Y^T L^{-1} Y \tag{5.120}$$

which is valid for all matrices $Z, Y \in \Re^{n \times k}$ and $\Lambda = \Lambda^T > 0, \Lambda \in \Re^{n \times n}$ [125], then it is possible to derive the following inequality from the right hand-side terms of Equation 5.119,

$$2e^T Ph \leqslant h^T H_n h + e^T P H_n^{-1} P e \tag{5.121}$$

Under Assumptions 5.3.1 to 5.3.5, $\dot{V}(e)$ in Equation 5.119 is bounded by,

$$\dot{V}(e) \leqslant -e^T Q_0 e + \varepsilon^0$$

$$= -e^T P^{1/2} \left(P^{-1/2} Q_0 P^{-1/2}\right) P^{1/2} e + \varepsilon^0$$

$$= -e^T P^{1/2} R_p P^{1/2} e + \varepsilon^0 \tag{5.122}$$

$$\leqslant -\lambda_{\min}(R_p) e^T P e + \varepsilon^0$$

$$= -\lambda_{\min}(R_p) V(e) + \varepsilon^0$$

where

$$R_p = P^{-1/2} Q_0 P^{-1/2} \tag{5.123}$$

Integrating Equation 5.122, the following expression is obtained,

$$V(e(t)) \leqslant V(e(0)) \exp(-\lambda_{\min}(R_P)t) + \varepsilon^0 \int_0^t \exp(-\lambda_{\min}(R_P)(t-\tau)) d\tau \tag{5.124}$$

which can be explicitly evaluated as,

$$V(e(t)) \leqslant V(e(0)) \exp(-\lambda_{\min}(R_P)t) + \frac{\varepsilon^0}{\lambda_{\min}(R_P)} (1 - \exp(-\lambda_{\min}(R_P)t)) \tag{5.125}$$

Consider the following inequality, which arises from Equation 5.118:

$$V(e(t)) \geqslant \lambda_{\min}(P) \|e(t)\|^2 \tag{5.126}$$

Then we have:

$$\|e(t)\| \leq \sqrt{\frac{V(e(t))}{\lambda_{\min}(P)}} \tag{5.127}$$

so that

$$\sup \|e(t)\| \leq \sqrt{\frac{\sup V(e(t))}{\lambda_{\min}(P)}} \tag{5.128}$$

But from Equation 5.125, due to the first order exponential nature of its right-hand side, we have:

$$\sup V(e(t)) \leq \max \left(V(e(0)), \frac{\varepsilon^0}{\lambda_{\min}(R_p)} \right) \equiv \eta \tag{5.129}$$

Therefore,

$$\sup \|e(t)\| \leq \sqrt{\frac{\eta}{\lambda_{\min}(P)}} \tag{5.130}$$

so that the expression given by Equation 5.114 has been obtained. □

Remark 5.3.8. This theorem provides a bound for the supremum (or least upper bound) of the norm of the modelling error e, which ultimately depends on the structure of the dynamic neural network (particularly its state dimension N), the activation function σ, the parameters of the dynamic neural network (matrices β, ω, γ), and some properties of the system being modelled.

5.4 Summary

This chapter has presented some important results on the approximation ability of multilayer perceptrons and dynamic neural networks. The ability of dynamic neural networks to approximate nonlinear autonomous systems has been studied and the importance of the hidden units in the approximation ability of the network has been noticed. This analysis was extended to non-autonomous nonlinear systems. Furthermore, a bound has been calculated for the norm of the approximation error for the case of non-autonomous systems.

CHAPTER 6
FEEDBACK LINEARISATION USING DYNAMIC NEURAL NETWORKS

The practical application of feedback linearising techniques has been limited by the requirement for a dynamic model of the plant. This requirement has constrained considerably the number of successful applications. The use of neural networks for feedback linearisation is a relatively recently introduced practice and has the potential to increase the number of applications, as neural network models can be calculated from measured data from the plant, so avoiding the need for deriving a physical model of a possibly complex process.

A dynamic neural network model can be trained to approximate the dynamic behaviour of a plant over an operating region, and it can be used to compute the linearising feedback law, in such a way that the plant exhibits an approximate linear behaviour from the the new inputs to its outputs [126]. Previous work in this field has demonstrated how a feedforward network can carry out linearisation [37] and how dynamic neural networks can be used as the basis for the design of robust control architectures [39]. Dynamic neural networks have also been used to input–output linearise SISO processes within the inverse model control framework [40, 41]. Input–output linearisation techniques using dynamic neural networks [42] and CMAC neural models [43] have also been proposed for SISO systems. Discrete time models have been used for feedback linearisation in chemical processes [44] and several SISO schemes for feedback linearisation using dynamic models parameterised by feedforward neural networks have been suggested [38]. In recent work, adaptive control techniques have been developed based on DNNs for induction motor control [45].

In this chapter, dynamic neural networks of the form presented in Chapter 4 are used to model nonlinear dynamic systems with multiple inputs and multiple outputs. A feedback linearising-decoupling law described in Chapter 3 is designed based on the neural model and then applied to the original plant. The analysis is first formulated for systems affine in the control input and later extended to general nonlinear systems. Conditions for the closed loop stability of the scheme are given and the effects of the approximation error are considered.

6.1 Approximate Input–Output Linearisation of Control Affine Systems

A dynamic neural network can be used to approximately input–output linearise and decouple a control affine nonlinear system. If the network successfully approximates the system following the identification procedure presented in Chapter 4, a linearising control law can be designed based on the dynamic neural network model and applied to the system achieving an approximate linearisation and decoupling. With the feedback linearisation laws employed, it is possible to choose the poles of the linearised system. An external control loop, which consists of multiple proportional+integral (PI) controllers, is designed in order to achieve typical specifications for each output such as time constants, zero steady-state error for constant setpoints and limited overshoot.

The following corollary, which is derived from Theorem 3.1.2, introduces the required transformations to achieve the input–output linearisation-decoupling of a dynamic neural network.

Corollary 6.1.1. *Given the dynamic neural network*

$$\dot{x}_n = -\beta x_n + \omega \sigma(x_n) + \gamma u \tag{6.1}$$
$$y_n = C_n x_n \tag{6.2}$$

where x_n are coordinates on $\Re^N$, $\beta \in \Re^{N \times N}$, $\omega \in \Re^{N \times N}$, $\gamma \in \Re^{N \times p}$, $u \in \Re^p$, $\sigma(x) = [\sigma(x_1), \ldots, \sigma(x_N)]^T$ and

$$C_n = \begin{bmatrix} I_{p \times p} & \emptyset_{p \times (N-p)} \end{bmatrix} \tag{6.3}$$

For arbitrary values $\hat{\lambda}_{ik}$ ($i = 1, \ldots, p$ and $k = 0, \ldots, r_i$) a state feedback law with

$$u = P(x_n) + Q(x_n)v \tag{6.4}$$

where

$$\begin{aligned} P(x_n) &= -A(x_n)^{-1} B(x_n) \\ Q(x_n) &= A(x_n)^{-1} \end{aligned} \tag{6.5}$$

with

$$A(x_n) = \begin{bmatrix} \hat{\lambda}_{1r_1} L_{g_1} L_f^{r_1-1} x_{n_1} & \cdots & \hat{\lambda}_{1r_1} L_{g_p} L_f^{r_1-1} x_{n_1} \\ \vdots & \ddots & \vdots \\ \hat{\lambda}_{pr_p} L_{g_1} L_f^{r_p-1} x_{n_p} & \cdots & \hat{\lambda}_{pr_p} L_{g_p} L_f^{r_p-1} x_{n_p} \end{bmatrix}_{p \times p} \tag{6.6}$$

and

6.1 Approximate Input–Output Linearisation of Control Affine Systems

$$B(x_n) = \begin{bmatrix} \sum_{k=0}^{r_1} \hat{\lambda}_{1k} L_f^k x_{n_1} \\ \vdots \\ \sum_{k=0}^{r_p} \hat{\lambda}_{pk} L_f^k x_{n_p} \end{bmatrix}_{p \times 1} \tag{6.7}$$

where the $\hat{\lambda}_{ik}$s are scalar design parameters and r_i is the relative degree of the ith output y_{n_i}, produces when applied to a system described by Equation 6.1, a linearised-decoupled system that obeys

$$\sum_{k=0}^{r_i} \hat{\lambda}_{ik} \frac{d^k y_{n_i}}{dt^k} = v_i, \quad i = 1 \ldots p \tag{6.8}$$

if the relative vector is well defined, $A(x_n)$ is invertible and

$$\det \left[\mathrm{diag} \left(\hat{\lambda}_{1r_1}, \hat{\lambda}_{2r_2}, \ldots, \hat{\lambda}_{pr_p} \right) \right] \neq 0 \tag{6.9}$$

Proof. The proof of this corollary follows from Theorem 3.1.2 by simply replacing the general control affine model given by Equation 3.2 with the dynamic neural network described by Equation 6.1. □

The following proposition shows that this control law produces approximately the same effect on the nonlinear system modelled by the dynamic neural network.

Proposition 6.1.1. *Consider the general nonlinear control affine system*

$$\dot{x} = f(x) + g(x)u \tag{6.10}$$
$$y = h(x) \tag{6.11}$$

where the state $x \in \Re^n$, and the dynamic neural network

$$\dot{x}_n = -\beta x_n + \omega \sigma(x_n) + \gamma u \tag{6.12}$$
$$y_n = C_n x_n \tag{6.13}$$

with $x_n \in \Re^N$,

$$C_n = \begin{bmatrix} I_{p \times p} & 0_{p \times (N-p)} \end{bmatrix} \tag{6.14}$$

both with p inputs, p outputs. Suppose that the dynamic neural network has vector relative degree $\{r_1, \ldots, r_p\}$. Suppose also that the parameter matrices β, ω and γ have been chosen so that the network approximates the system given by Equation 6.10 over a region of interest. If the linearising-decoupling law 6.4 designed for the dynamic neural network 6.12 under Corollary 6.1.1 is applied to system 6.10, then the dynamics of the plant 6.10 are approximately input–output linearised and decoupled, and obey the following differential equations:

124   6. Feedback Linearisation Using DNNs

$$\sum_{k=0}^{r_i} \hat{\lambda}_{ik} \frac{d^k y_i}{dt^k} = v_i + \sum_{k=0}^{r_i} \hat{\lambda}_{ik} \frac{d^k e_i}{dt^k}, \quad i = 1, \ldots, p \tag{6.15}$$

where e_i is the model error corresponding to the ith output:

$$e_i = y_i - y_{n_i} \tag{6.16}$$

Proof. Consider the dynamic neural network in Equation 6.12. According to Corollary 6.1.1, when the linearising-decoupling law in Equation 6.4 is applied to this network, the result is that the network is input–output linearised and decoupled as in Equation 6.8, which is repeated here:

$$\sum_{k=0}^{r_i} \hat{\lambda}_{ik} \frac{d^k y_{n_i}}{dt^k} = v_i \tag{6.17}$$

Using the expression for the model error in Equation 6.16, we find that the ith output of the network is given by:

$$y_{n_i} = y_i - e_i \tag{6.18}$$

Replacing Equation 6.18 into 6.17, we find:

$$\sum_{k=0}^{r_i} \hat{\lambda}_{ik} \frac{d^k (y_i - e_i)}{dt^k} = v_i \tag{6.19}$$

so that:

$$\sum_{k=0}^{r_i} \hat{\lambda}_{ik} \frac{d^k y_i}{dt^k} = v_i + \sum_{k=0}^{r_i} \hat{\lambda}_{ik} \frac{d^k e_i}{dt^k} \tag{6.20}$$

□

Remark 6.1.1. Notice that the ith output y_i obeys a linear differential equation with inputs v_i and e_i. The model error e_i will in general depend in a complex manner on the other inputs and also on the states of the system and of the network.

Remark 6.1.2. In the Laplace domain, Equation 6.20 can be expressed as follows:

$$y_i(s) = \frac{1}{\lambda_{ir_i} s^r + \cdots + \lambda_{i0}} v(s) + e_i(s) \tag{6.21}$$

so that the model error can be seen as an output disturbance on output $y_i(s)$.

Remark 6.1.3. It is possible to ensure that the model error e_i remains bounded by appropriate training as seen in Chapters 4 and 5, so that its effect on the linearising and decoupling action remains small within a region of the input–output space.

Remark 6.1.4. Notice that the state dimension of the network (N) is not required to be the same as the state dimension of the system (n).

6.1.1 Approximate input–output linearisation-decoupling and external control

In this section we describe how to approximately input–output linearise and decouple the plant using the method given in Proposition 6.1.1 and then how to apply external control to the linearised and decoupled system.

If a dynamic neural network of the form shown in Equation 6.1 is used as a dynamic model of the system in Equation 6.10; then the control law is synthesised based on the neural model. Once the network is trained using the method described in Section 4.6, its vector relative degree at a point x_o is obtained by checking that

$$L_{g_j} L_f^k h_i(x) = 0, j = 1, \ldots, p, k = 0, \ldots, r_i - 2 \tag{6.22}$$

for each output and verifying that the characteristic matrix $C(x)$, which is given by Equation 6.23, is non-singular:

$$C(x) = \begin{bmatrix} \sum_{i=1}^{N} \gamma_{i1} \frac{\partial L_f^{r_1-1} x_1}{\partial x_i} & \cdots & \sum_{i=1}^{N} \gamma_{ip} \frac{\partial L_f^{r_1-1} x_1}{\partial x_i} \\ \vdots & \ddots & \vdots \\ \sum_{i=1}^{N} \gamma_{i1} \frac{\partial L_f^{r_p-1} x_p}{\partial x_i} & \cdots & \sum_{i=1}^{N} \gamma_{ip} \frac{\partial L_f^{r_p-1} x_p}{\partial x_i} \end{bmatrix}_{p \times p} \tag{6.23}$$

For the specific cases of $r_i = 1, 2$ $(i = 1, \ldots, p)$, the expressions for $A(x)$ and $B(x)$ given in Equation 6.5 can be expressed as follows:

$$A(x) = \begin{bmatrix} \hat{\lambda}_{1r_1} v_{11} & \cdots & \hat{\lambda}_{1r_1} v_{1p} \\ \vdots & \ddots & \vdots \\ \hat{\lambda}_{pr_p} v_{p1} & \cdots & \hat{\lambda}_{pr_p} v_{pp} \end{bmatrix}_{p \times p} \quad B(x) = \begin{bmatrix} \sum_{k=0}^{r_1} \hat{\lambda}_{1k} q_{1k} \\ \vdots \\ \sum_{k=0}^{r_p} \hat{\lambda}_{pk} q_{pk} \end{bmatrix}_{p \times 1} \tag{6.24}$$

given the ith row of matrix $A(x)$, corresponding to the ith output, if $r_i = 1$, then

$$v_{ij} = \gamma_{ij}, \quad j = 1 \ldots p \tag{6.25}$$

and if $r_i = 2$, then

$$v_{ij} = \sum_{k=1}^{N} \gamma_{kj} \left[-\beta_{ik} + \omega_{ki} \left(1 - \sigma^2(x_i)\right) \right], \quad j = 1 \ldots p \tag{6.26}$$

with

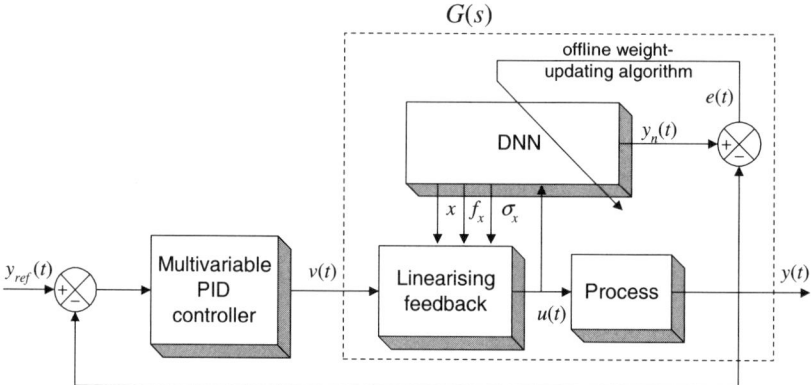

Fig. 6.1. Feedback linearisation using a DNN including an external control loop

$$q_{i0} = x_i \tag{6.27}$$

$$q_{i1} = f_i = -\beta_{ii}x_i + \sum_{j=1}^{N} \omega_{ij}\sigma(x_j) \tag{6.28}$$

$$q_{i2} = \sum_{k=1}^{N}\left(-\beta_{kk}x_k + \sum_{j=1}^{N}\omega_{kj}\sigma(x_j)\right)\left(-\beta_{ik} + \omega_{ik}\left(1 - \sigma^2(x_k)\right)\right) \tag{6.29}$$

$$\tag{6.30}$$

$\beta_{ij} = \beta_i$ if $i = j$, 0 in other cases. With γ_{ij}, β_{ij} and ω_{ij} taken from the identified dynamic neural network model.

These formulas are valid for the activation function

$$\sigma(x) = \tanh(x) = \frac{e^x - e^{-x}}{e^x + e^{-x}} \tag{6.31}$$

which satisfies

$$\sigma'(x_i) = \frac{d\sigma(x_i)}{dx_i} = (1 - \sigma^2(x_i)); \ x_i \in \Re \tag{6.32}$$

The smoothness, boundedness and monotony of the hyperbolic tangent function (see Equation 6.31) as the activation function of the neural network, make it a suitable selection. In addition, the property of representing its derivatives as functions of the original form remarkably simplifies the linearising feedback law. The parameters $\hat{\lambda}_{ik}$ are adjustable and can be chosen to accomplish design objectives, such as stability and speed of response of the $v - y$ system, which is illustrated in Figure 6.1. The dynamics can be further shaped at a second stage using a controller across the $v - y$ system.

6.1 Approximate Input–Output Linearisation of Control Affine Systems

Multivariable PI Control. A conventional multivariable PI controller may be used to close the loops of the linearised system. The control law applied to the $v - y$ system can be represented by

$$v_i = K_{p_i}\left(y_{ref_i}(t) - y_i(t)\right) + K_{I_i} \int_{t_0}^{t_1} \left(y_{ref_i}(t) - y_i(t)\right) dt, \quad i = 1, \ldots, p \quad (6.33)$$

where y_{ref_i} and y_i, $i = 1, \ldots, p$, are the setpoint and process outputs signals respectively. K_{p_i} and K_{I_i} are the controller gains for the ith loop, which are adjusted to achieve satisfactory responses in the output of the process. Figure 6.1 illustrates the proposed control diagram with an internal state feedback and an external linear controller. The combination of external PI controllers and input–output linearising feedback has been termed in the literature as *globally linearising control* [63].

The design procedure of the controller can be summarised in four steps:

1. Neural system identification of the process with the use of an appropriate training data set.
2. Computation of the input–output linearising and decoupling feedback.
3. Selection of suitable $\hat{\lambda}_{ik}$s achieving desired dynamics and stability on the $v - y$ system.
4. Design of a conventional linear PI controller across the previously linearised system.

Disturbance rejection. Since the feedback law in Equation 6.4 does not provide integral action, the outputs may exhibit steady state error for a constant reference under proportional only control. The multivariable proportional-integral control law in Equation 6.33 eliminates the effect of the model error e_i on the output. In order to evaluate the effect of including the multivariable PI controller on the plant response, the PI control law in Equation 6.33 is applied to the linearised-decoupled system in Equation 6.15, which results in:

$$\sum_{k=0}^{r_i} \hat{\lambda}_{ik} \frac{d^k y_i}{dt^k} = K_{p_i}\left(y_{ref_i}(t) - y_i(t)\right)$$
$$+ K_{I_i} \int_{t_0}^{t_1} \left(y_{ref_i}(t) - y_i(t)\right) dt \quad (6.34)$$
$$+ \sum_{k=0}^{r_i} \hat{\lambda}_{ik} \frac{d^k e_i}{dt^k}$$

Differentiating Equation 6.34 results in

$$\sum_{k=1}^{r_i+1} \hat{\lambda}_{i,k-1} \frac{d^k y_i}{dt^k} = K_{p_i}\left(\dot{y}_{ref_i} - \dot{y}_i\right) + K_{I_i}\left(y_{ref_i} - y_i\right)$$
$$+ \sum_{k=1}^{r_i+1} \hat{\lambda}_{i,k-1} \frac{d^k e_i}{dt^k} \quad (6.35)$$

and collecting similar terms yields

$$K_{I_i}y_i + \left(\hat{\lambda}_{i1} + K_{p_i}\right)\dot{y}_i + \sum_{k=2}^{r_i+1}\hat{\lambda}_{i,k-1}\frac{d^k y_i}{dt^k} = K_{p_i}\dot{y}_{ref_i} + K_{I_i}y_{ref_i} \\ + \sum_{k=1}^{r_i+1}\hat{\lambda}_{i,k-1}\frac{d^k e_i}{dt^k} \quad (6.36)$$

In the Laplace domain, Equation 6.36 is given by:

$$Y_i(s) = \frac{K_{I_i}+K_{p_i}s}{K_{I_i}+(\hat{\lambda}_{i1}+K_{p_i})s+\sum_{k=2}^{r_i+1}\hat{\lambda}_{i,k-1}s^k}Y_{ref_i}(s) \\ + \frac{\sum_{k=1}^{r_i+1}\hat{\lambda}_{ik}s^k}{K_{I_i}+(\hat{\lambda}_{i1}+K_{p_i})s+\sum_{k=2}^{r_i+1}\hat{\lambda}_{i,k-1}s^k}E_i(s) \quad (6.37)$$

and the control error of the ith output is given by

$$E_{c_i}(s) = Y_{ref_i}(s) - Y_i(s) \\ = \left[1 - \frac{K_{I_i}+K_{p_i}s}{K_{I_i}+(\hat{\lambda}_{i1}+K_{p_i})s+\sum_{k=2}^{r_i+1}\hat{\lambda}_{i,k-1}s^k}\right]Y_{ref_i}(s) \\ - \frac{\sum_{k=1}^{r_i+1}\hat{\lambda}_{i,k-1}s^k}{K_{I_i}+(\hat{\lambda}_{i1}+K_{p_i})s+\sum_{k=2}^{r_i+1}\hat{\lambda}_{i,k-1}s^k}E_i(s) \quad (6.38)$$

For simplicity of the analysis, assume that both the reference signal $y_{ref_i}(t)$ and the model error $e_i(t)$ are constant. Then, using the final value theorem [127], it is easy to show that the control error in steady state is zero:

$$e_{c_i}(\infty) = \lim_{t\to\infty} y_{ref_i} - y_i(t) = 0 \quad (6.39)$$

Thus, the external PI loop cancels the offset introduced by a constant model error. Given that the model error e_i is equivalent to a measurable disturbance, it is also possible to add a lead/lag controller to dynamically reject the model error. Disturbance rejection is largely documented in the linear control literature [128, 129]. Suitable training will guarantee that the dynamic neural network is sufficiently accurate to maintain the model error $e_i(t)$ within desired bounds as seen in Chapters 4 and 5.

6.1.2 Stability analysis

The stability of closed loop systems governed by a given control law does not guarantee the achievement of design objectives, but it does represent an important condition to prove. In previous chapters, the stability conditions for dynamic neural networks and input–output linearised systems were given.

6.1 Approximate Input–Output Linearisation of Control Affine Systems

Based on Proposition 3.1.5, the following proposition states the stability conditions for approximate input–output linearised systems using dynamic neural networks.

Proposition 6.1.2. *Consider the closed loop response $x(t, x(0))$ of a system described by*

$$\dot{x} = f(x) + g(x)u \tag{6.40}$$

under the linearising-decoupling control law

$$u = P(x_n) + Q(x_n)v \tag{6.41}$$

designed for the dynamic neural network described by Equation 6.1, with $P(x_n)$ and $Q(x_n)$ given by Equation 6.5, where $x_n(t, x_n(0))$ is the solution of Equation 6.1 under the same control law 6.41. If the following sufficient conditions are satisfied:

1. *The dynamic neural network described by Equation 6.1 is input-to-state stable.*
2. *The autonomous system $\dot{x} = f(x)$ is asymptotically stable, the the external input v is bounded and the initial conditions are such that for each $\varepsilon > 0$ there exists $\delta_1 > 0$ and $K > 0$ such that:*

$$\|x(0)\| < \delta_1$$
$$\|P(x_n) + Q(x_n)v(t)\| < K$$

then the solution $x(t, x(0))$ is bounded:

$$\|x(t, x(0))\| < \varepsilon, \quad \text{for all } t \geq 0$$

Proof. When the linearising feedback given by Equation 6.41 is applied to the system in Equation 6.40, we have:

$$\dot{x} = f(x) + g(x)u = f(x) + g(x)\left(P(x_n) + Q(x_n)v\right) \tag{6.42}$$

If the dynamic neural network is input-to-state stable (see Theorem 4.5.3), then its state $x_n(t, x_n(0))$ and so $P(x_n)$ and $Q(x_n)$ are bounded as they are smooth, such that:

$$\|P(x_n)\| < K_1$$
$$\|Q(x_n)\| < K_2$$

for some K_1 and K_2, $0 < K_1 < \infty$ and $0 < K_2 < \infty$. If, furthermore, the external input v is bounded such that $\|v\| < K_3$ for some K_3, $0 < K_3 < \infty$, then:

$$||u|| = ||P(x_n) + Q(x_n)v|| < ||P(x_n)|| + ||Q(x_n)||||v||$$
$$< K_1 + K_2 K_3$$

so that u is bounded as well. Since the autonomous system $\dot{x} = f(x)$ is asymptotically stable, then by Proposition 3.1.7, for each $\varepsilon > 0$ there exists $\delta_1 > 0$ and $K > 0$ such that if $||x(0)|| < \delta_1$ and $||u(t)|| < K$ for all $t \geq 0$, then the solution $x(t, x(0))$ of the system in Equation 6.40 is bounded:

$$||x(t, x(0))|| < \varepsilon, \quad \text{for all } t \geq 0 \tag{6.43}$$

□

Stability under external PI control. The stability of the closed loop system including the external multivariable PI controller depends on the selection of suitable design parameters $\hat{\lambda}_{ik}$s in Equations 6.6 and 6.7 and on the selection of suitable controller gains K_{p_i} and K_{I_i}. The roots of the characteristic polynomials $P_i(s)$, $i = 1, \ldots, p$ in Equation 6.37:

$$P_i(s) = K_{I_i} + \left(\hat{\lambda}_{i1} + K_{p_i}\right)s + \sum_{k=2}^{r_i+1} \hat{\lambda}_{i,k-1} s^k \tag{6.44}$$

determines the position of the poles in each input–output channel $v_i - y_i$, $i = 1, \ldots, p$, thus the stability of the external closed loop system. An additional condition is, once more, the boundedness of both the external input v_i and the modelling error of the network e_i for each output $i = 1, \ldots, p$.

6.2 Approximate Input–Output Linearisation for General Nonlinear Systems

The control affine neural network in Equation 6.1 can approximate not only control affine systems but also general nonlinear systems of the form $\dot{x} = f(x, u)$. Thus, the dynamic neural network model and its associated relative degree vector constitute the basis for designing the linearising-decoupling law for general nonlinear systems as was done in the control affine case.

In general, the solution to the input–output linearisation law for general nonlinear systems cannot be found analytically. For this reason, the classical design strategy is based on the development of an extended affine model that is directly coupled to a new manipulated input [73]. The inherent approximation capability of dynamic neural networks makes it possible for the same control strategy designed for affine systems to approximately linearise and decouple non-affine systems. The following proposition shows how the general nonlinear system is approximately linearised and decoupled based on a dynamic neural network model.

Proposition 6.2.1. *Consider the general nonlinear system*

$$\dot{x} = f(x,u) \\ y = h(x) \tag{6.45}$$

where state $x \in \Re^n$ is the state vector, $y \in \Re^p$ is the output vector and $u \in \Re^p$ is the input vector, and consider also the dynamic neural network

$$\dot{x}_n = -\beta x_n + \omega \sigma(x_n) + \gamma u \\ y_n = C_n x_n \tag{6.46}$$

where $x_n \in \Re^N$ is the network state vector, $y_n \in \Re^p$ is the network output vector, $u \in \Re^p$ is the input vector, the parameter matrices β, ω and γ have been chosen so that the network approximates the system given by Equation 6.45 over a region of interest, and

$$C_n = \begin{bmatrix} I_{p \times p} & \emptyset_{p \times (N-p)} \end{bmatrix} \tag{6.47}$$

Suppose that the network has a vector relative degree $\{r_1, \ldots, r_p\}$. If the linearising-decoupling control signal given by Equation 6.4 designed for the dynamic neural network under Corollary 6.1.1 is applied to the system described by Equation 6.45, the dynamics of the plant are approximately linearised-decoupled such that the ith output obeys the following differential equation:

$$\sum_{k=0}^{r_i} \hat{\lambda}_{ik} \frac{d^k y_i}{dt^k} = v_i + \sum_{k=0}^{r_i} \hat{\lambda}_{ik} \frac{d^k e_i}{dt^k} \tag{6.48}$$

where e_i is the model error corresponding to the ith output:

$$e_i = y_i - y_{n_i} \tag{6.49}$$

Proof. The proof of this proposition is identical to the proof of Proposition 6.1.1. □

The control algorithm for the general nonlinear case is exactly the same as the control affine case discussed in Section 6.1. After the system given by Equation 6.45 is approximately linearised and decoupled, the multivariable PI control law described by Equation 6.33 may be applied. This scheme also provides the same disturbance rejection capabilities as in the affine case, which is illustrated in Figure 6.3.

6.3 Related Work

An indirect control scheme for nonlinear systems described in Reference [130] was proposed using a similar configuration of dynamic networks, where the

132   6. Feedback Linearisation Using DNNs

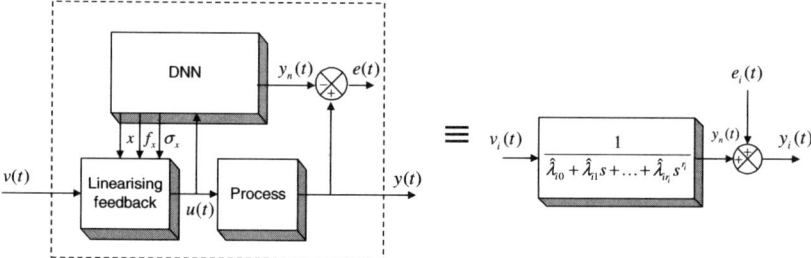

Fig. 6.2. Equivalent block diagram for the linearised-decoupled system. On the right-hand side there is one independent block for every output $i = 1, \ldots, p$. The modelling error is considered as a measured output disturbance

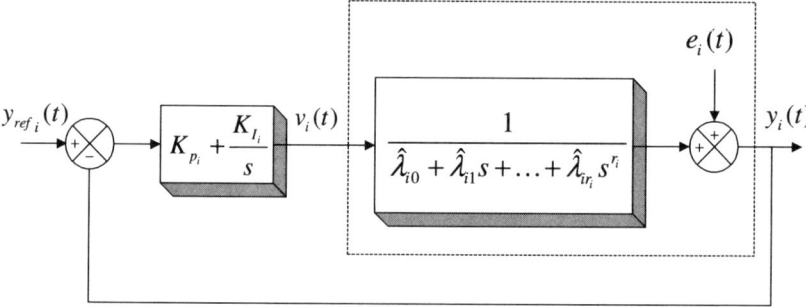

Fig. 6.3. MIMO globally linearising control structure. As a result of the linearising-decoupling feedback, there is one independent loop for each output $i = 1, \ldots, p$

restrictions on the system are stronger than with the methods presented in this chapter and the approach is limited to control affine systems. In the approach presented in Reference [38], the system was identified with dynamic models parameterised by static multilayer networks. The technique was proposed for SISO systems and is not easily expandable to multivariable systems. The method described in Reference [55] a dynamic model based on static neural networks is used to input–output linearise a nonlinear system and then predictive control is externally applied to the linearised system.

A method described in Reference [40] focused on the use of a two-neuron dynamic neural network to input–output linearise a multivariable process within an inverse model control framework. The restricted number of neurons and the unitary relative vector of the control affine neural model compromise the approximation capability of the network. Considering only unitary relative degrees for the outputs drastically reduces the linearisation effect on systems with higher vector relative degree. Similarly, in a linearising control strategy developed for induction motors [45], the relative degree of the network is also fixed, such that $r_i = 1$, $i = 1, \ldots, p$, which reduces the possibility of reproducing the behaviour of systems with higher relative degrees. This difficulty is overcome by using a large number of neurons to compensate the low relative degree structure of the network. Nevertheless, the use of an excessive number of neurons may increase significantly the number of parameters and computational cost.

In the approach presented in this chapter, the vector relative degree of the network depends on the values of the network parameters and it may take values $r_i \geq 1$, $i = 1, \ldots, p$. Furthermore, if the vector relative degree of the plant is known *a priori*, it is possible to include constraints on the parameter values as part of the optimisation problem associated with training, such that the resulting vector relative degree of the network is similar to the vector relative degree of the plant.

6.4 Summary

This chapter has presented a technique for input–output linearisation and decoupling based on dynamic neural networks. The identification procedure described in Chapter 4 produces a state space model that approximates the dynamics of the plant and which is used as the basis for input–output linearisation. The resulting linearising transformations are then applied to the actual plant, using the state vector of the dynamic neural network model to compute these transformations. The study, which includes stability and disturbance rejection analysis, shows that the method has the potential to succeed in approximately linearising and decoupling multivariable nonlinear systems. The analysis is first presented for control affine systems and later extended to general nonlinear systems.

The linearised decoupled system is then immersed in a closed loop employing proportional+integral (PI) controllers in order to achieve design specifications, which may include desired time constants and zero steady state errors for constant reference signals. The use of integral action has the added benefit of rejecting the output disturbance introduced by the model error.

CHAPTER 7
CASE STUDIES

This chapter presents three case studies that illustrate the identification and control strategies presented in this book. First, in Section 7.1, a dynamic neural network model is trained based on measurements taken from a pressure pilot plant, and the effects of correct initialisation of the hidden states of the network both during training and during usage, are highlighted. The globally linearising control strategy discussed in Chapter 6 and based on the identified dynamic neural network model was applied on-line to the pressure plant. Secondly, Section 7.2 presents a case study involving the approximate feedback linearisation of a simulated single link manipulator. Finally, Section 7.3 presents a case study involving identification and approximate feedback linearisation and decoupling of a simulated evaporator process.

7.1 The Pressure Pilot Plant

7.1.1 Description of the plant

The pressure pilot plant[1] used in this case study [103] is illustrated in Figure 7.1. It consists of a pressure vessel containing air and water. The air pressure is measured at the top of the vessel by means of a pressure transducer. A hydraulic pump is used to create a water flow that enters the vessel through an inlet pipe and so decreases the air volume, thus increasing its pressure. For a given pump rotation speed the system reaches an equilibrium point where no extra water enters the vessel. Furthermore, the direction of flow can be reversed such that the level decreases and so does the air pressure. The input signal, with a range of 0–10 V, is the voltage applied to the power amplifier that drives the DC motor that operates the hydraulic pump. The signals are sent and acquired by a supervisory PC via a Profibus network. The pressure signal ranges between 0 and 100 mbar. The sampling time used was 0.165 s.

---

[1] This plant is located at the Mechanical Engineering Department, ISEL, Lisbon Polytechnic Institute, Portugal

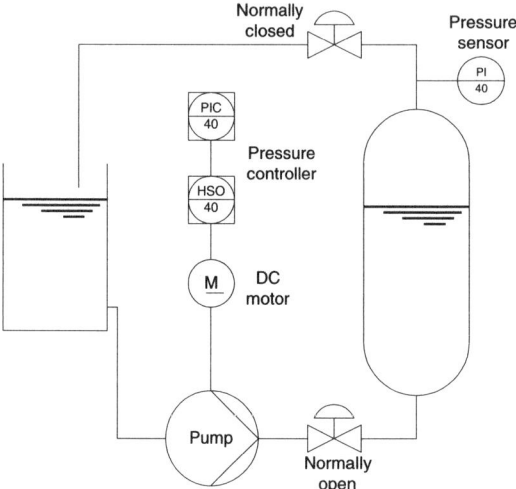

Fig. 7.1. Schematic diagram of the pilot plant

7.1.2 Identification results

Both the training and validation data sets, which are illustrated in Figure 7.2, were obtained experimentally and had 500 input–output samples each. The optimisation associated with training the dynamic neural network model was carried out using a quasi-Newton unconstrained algorithm implemented in function *fminunc*, which is part of the Optimisation Toolbox in MATLAB® version 6.1. This case study compares the training and validation performance of the models calculated with and without optimisation of the initial values of the hidden states.

The best model found had two states: an output state and a hidden state. Figures 7.3 and 7.4 show the training and validation trajectories, respectively, with optimised initial hidden states, as described in Sections 4.6. Figures 7.5 and 7.6 show the training and validation data, respectively, with the initial hidden state set to zero. Notice the differences between the results with and without optimised initial hidden state.

The values of the parameters when the initial hidden state was optimised were:

$$\beta = \begin{bmatrix} 0.7074 & 0 \\ 0 & 10.9863 \end{bmatrix} \tag{7.1}$$

$$\omega = \begin{bmatrix} -9.9086 & 23.0508 \\ 10.2314 & 6.8820 \end{bmatrix} \tag{7.2}$$

$$\gamma = \begin{bmatrix} 1.1016 \\ -0.4945 \end{bmatrix} \tag{7.3}$$

$$C_n = \begin{bmatrix} 1 & 0 \end{bmatrix} \tag{7.4}$$

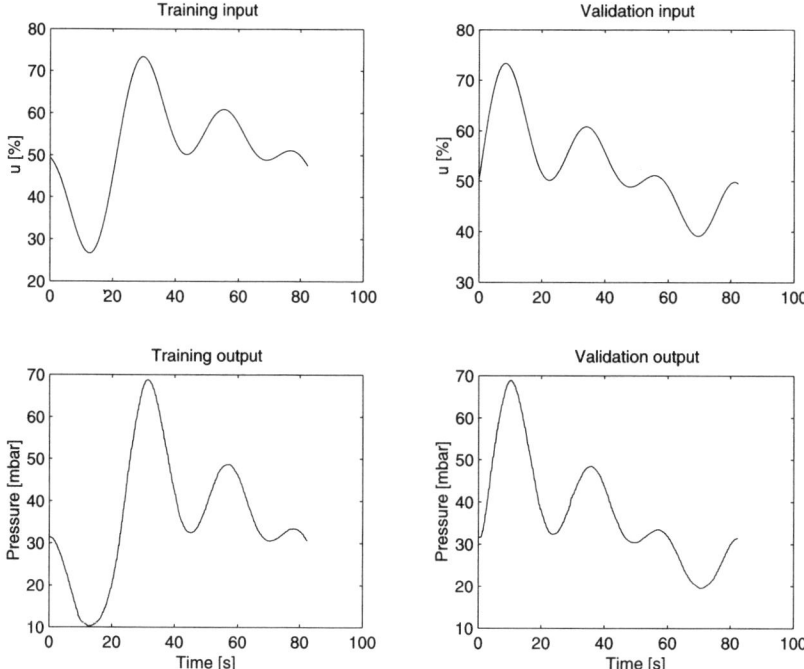

Fig. 7.2. Pressure plant. Training and validation data

7.1.3 Globally linearising control experimental results

The globally linearising control strategy described in Chapter 6 and illustrated in Figure 6.3 was implemented in SIMULINK® and applied to the real pilot process. The identified dynamic neural model has relative degree $r = 1$, and this relative degree was used to computed the feedback linearising law based on the model. The feedback law used was therefore as follows (see Example 3.1.7):

$$u = \frac{v - \lambda_0 x_1 - \lambda_1(-\beta_1 x_1 + \omega_{11}\sigma(x_1) + \omega_{12}\sigma(x_2))}{\lambda_1 \gamma_1} \quad (7.5)$$

where $\lambda_0 = 0.697$ and $\lambda_1 = 0.9853$ were computed such that the linearised system $v - y$ had the same static gain and dominant time constant of the Jacobian linearisation of the dynamic neural network model, β_1, ω_{11}, ω_{12} and γ_1 were taken from the identified dynamic neural network model. The PI controller parameters were $K_p = 1.5$ and $K_i = 1$ in all the experiments.

Four experiments were carried out. Figure 7.7 shows the trajectories for the globally linearising control strategy for a square wave reference, where v is the PI controller output and u is the process input. Figure 7.8 shows the trajectories for a square wave reference for the case of PI control only but

138   7. Case Studies

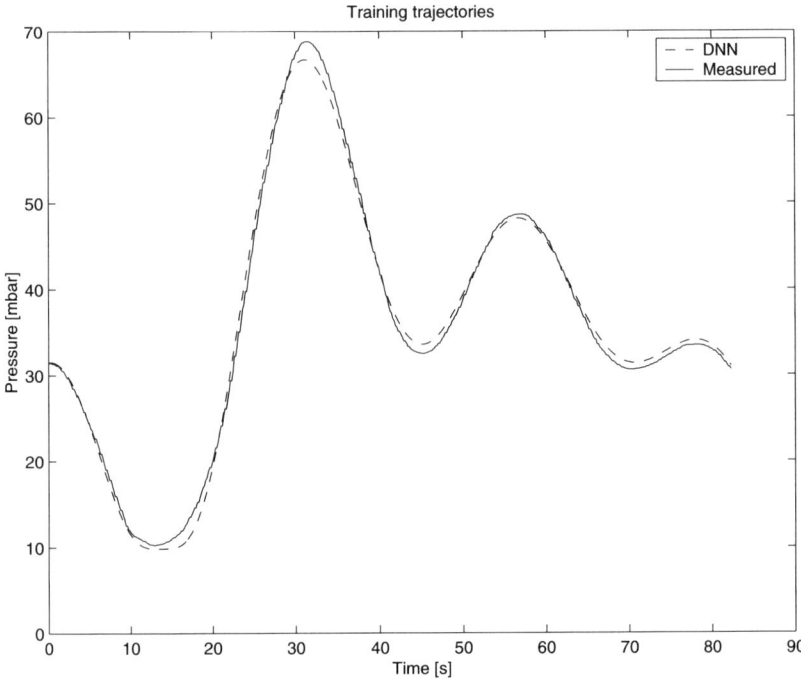

Fig. 7.3. Pressure plant. Training trajectories with optimised initial hidden state

no feedback linearisation. Figure 7.7 shows the trajectories for the globally linearising control strategy for a sinusoidal reference. Figure 7.8 shows the trajectories for a sinusoidal reference for the case of PI control only but no feedback linearisation. Notice the weaker asymmetry in the responses for the case with square wave reference with globally linearising control (compared with the corresponding case of PI control only) and also the smaller harmonic content of the y and v signals in the case with sinusoidal reference and globally linearising control (compared with the corresponding case with PI control only).

7.2 Single Link Manipulator

The globally linearising control strategy described in Chapter 6 and illustrated in Figure 6.3 was implemented in SIMULINK® and applied to the model of the single link manipulator described in Section 4.8 (see Equation 4.116). The identified dynamic neural model has relative degree $r = 1$ (see Equations 4.117 to 4.120 and Example 3.1.3). However, the zero dynamics of the identified model do not satisfy the sufficient conditions for asymptotic stability of the equilibrium $\eta = 0$, which were found in Example 3.1.5. Notice

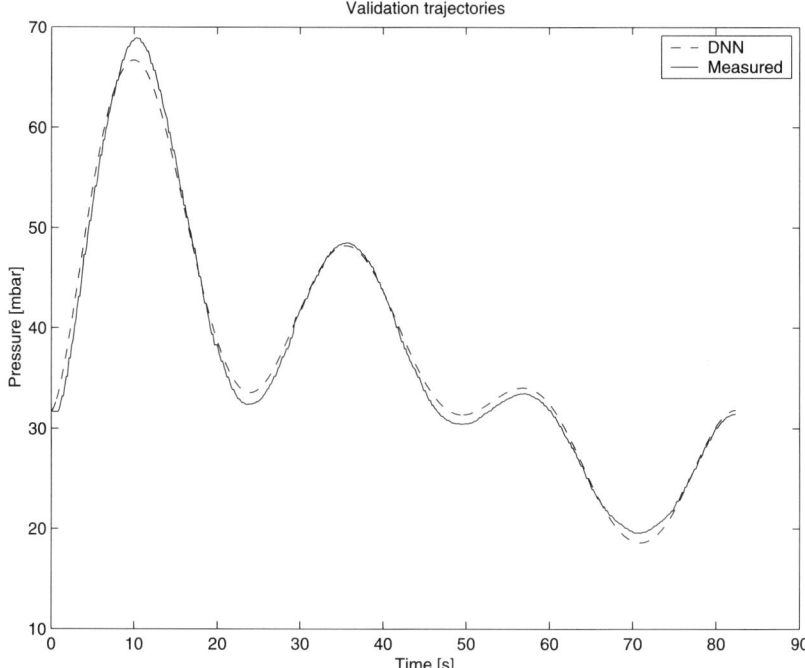

Fig. 7.4. Pressure plant. Validation trajectories with optimised initial hidden state

that the value of γ_1 is of small magnitude compared with the other parameters and that one of the conditions for having relative degree $r = 2$ is that $\gamma_1 = 0$ (see Example 3.1.3). Because of these facts, following Remark 3.1.4, better results were obtained assuming that the relative degree was $\rho = 2$, which corresponds to the relative degree of the physical model (see Equation 4.116), to compute the feedback linearising law based on the identified model. The feedback law used was therefore as follows (see Example 3.1.7):

$$u = \frac{v - \sum_{k=0}^{\rho} \lambda_k L_f^k h(x)}{\lambda_r L_g L_f^{\rho-1} h(x)} = \frac{v - \lambda_0 L_f^0[x_1] - \lambda_1 L_f^1[x_1] - \lambda_2 L_f^2[x_1]}{\lambda_2 L_g L_f^1[x_1]}$$

$$= \frac{v - \lambda_0 x_1 - \lambda_1(-\beta_1 x_1 + \omega_{11}\sigma(x_1) + \omega_{12}\sigma(x_2))}{D} \qquad (7.6)$$

$$- \frac{\lambda_2(-\beta_1 x_1 + \omega_{11}\sigma(x_1) + \omega_{12}\sigma(x_2))(-\beta_1 + \omega_{11}\sigma'(x_1))}{D}$$

$$- \frac{\lambda_2(-\beta_2 x_2 + \omega_{21}\sigma(x_1) + \omega_{22}\sigma(x_2))\omega_{12}\sigma'(x_2)}{D}$$

with

$$D = \lambda_2((-\beta_1 + \omega_{11}\sigma'(x_1)) + \gamma_2\omega_{12}\sigma'(x_2)) \qquad (7.7)$$

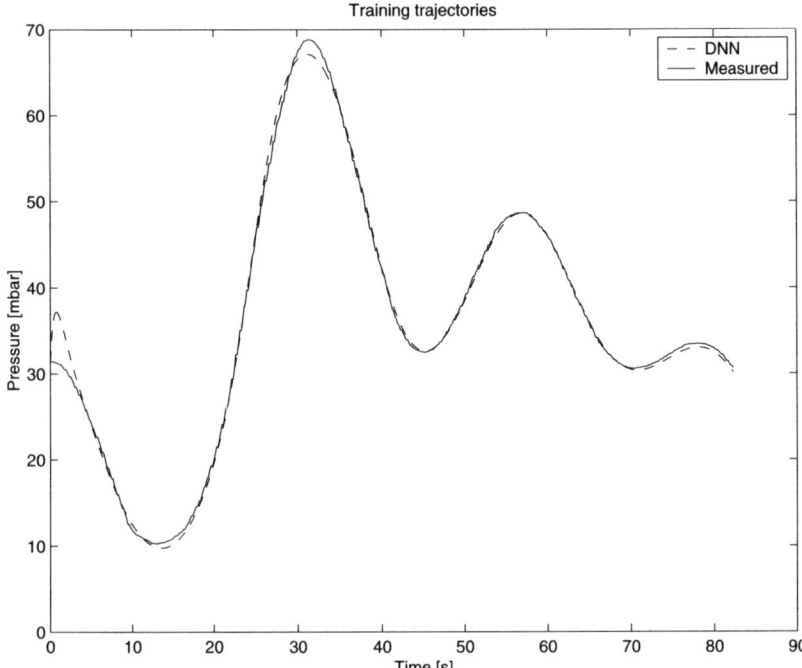

Fig. 7.5. Pressure plant. Training trajectories with zero initial hidden state

where $\lambda_0 = 10.1956$, $\lambda_1 = 3.0176$ and $\lambda_2 = 1$ were computed such that the linearised system $v - y$ had the same static gain and complex poles as the Jacobian linearisation of the dynamic neural network model, the parameters β, ω, and γ were taken from the identified dynamic neural network model given in Equations 4.117 to 4.119. The PI controller parameters used were $K_p = 20$ and $K_i = 10$.

The reference used in the simulations was a sinusoidal with frequency 0.2 Hz and amplitude 0.7 rad, offset by a step at $t = 0$ of magnitude 0.9 rad. Figure 7.11 shows the trajectories under globally linearising control. Notice the linearisation effect, since the input $u(t)$ contains significant harmonics, while the external input $v(t)$, which in this case is the controller output, and the manipulator output, do not show any noticeable harmonics. Figure 7.12 shows the resulting trajectories under PI control and no feedback linearisation. Notice that the input $u(t)$, which in this case is the controller output, contains significant harmonics.

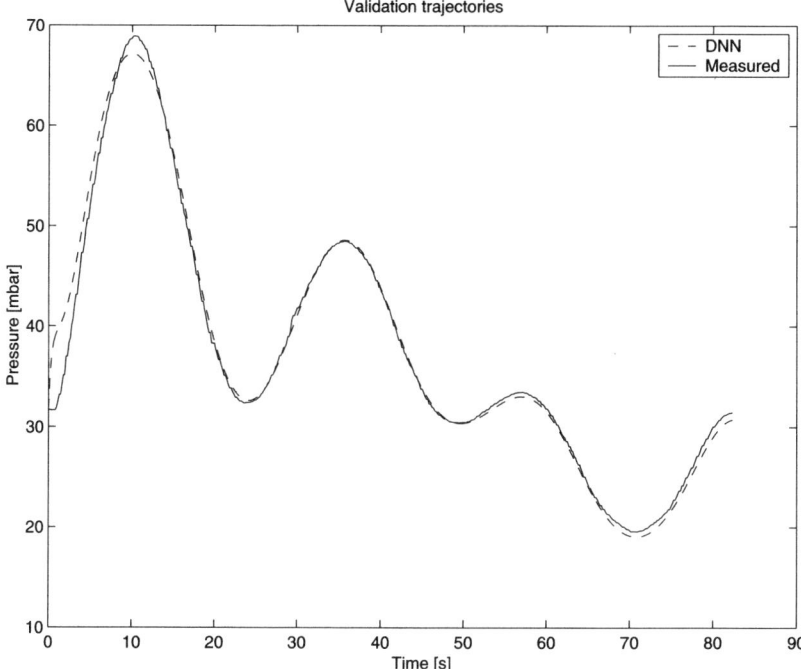

Fig. 7.6. Pressure plant. Validation trajectories with zero initial hidden state

7.3 Evaporator System

The concentration of dilute liquors obtained by evaporation of the solvent from a liquid mixture represents an important process in industries such as sugar mills, paper manufacture and food processing. The basic concept of evaporation is to use heat as a means of changing the composition of liquid mixtures by the vaporisation of one or more of its components. Pure solvent is often removed from a liquid that contains a non-volatile substance or solute, which is a component that does not vaporise.

The evaporation process uses a considerable amount of heat, so that effective process control schemes are necessary to ensure the overall system profitability, energy efficiency and an environment-friendly operation. The evaporator model employed in this case study represents a forced circulation evaporator [131]. Consider the diagram shown in Figure 7.13. The flow feed is mixed with the liquor and then pumped into a vertical heat exchanger. The steam is used to provide heat in the heat exchanger and then condenses on the outside of the tube walls. The heated mixture is then passed on to a separation vessel in which liquid and vapour are separated. The liquid is partly recirculated and partly drawn off as product. The vapour is condensed by cooling, with water often being used as the coolant agent.

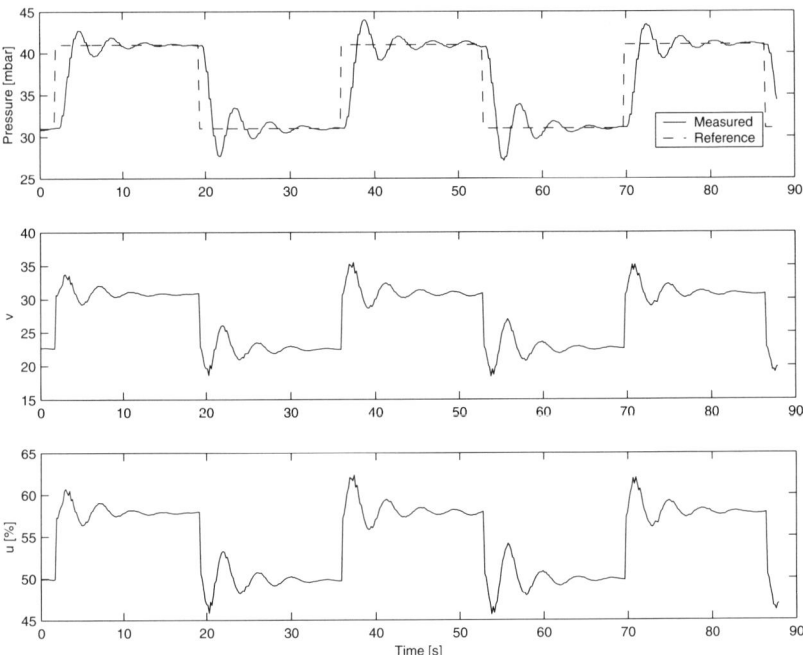

Fig. 7.7. Pressure plant. Globally linearising control trajectories for a square wave reference

Mathematical model. The mathematical model of the evaporator system was derived using mass and energy balance [132]. The differential and algebraic equations involved in the model are given below:

$$\frac{dL_2}{dt} = \frac{1}{\rho A}(F_1 - F_4 - F_2) \tag{7.8}$$

$$\frac{dX_2}{dt} = \frac{1}{M}(F_1 X_1 - F_2 X_2) \tag{7.9}$$

$$\frac{dP_2}{dt} = \frac{1}{C}(F_4 - F_5) \tag{7.10}$$

$$T_2 = 0.5616 P_2 + 0.3126 X_2 + 48.43 \tag{7.11}$$

$$T_3 = 0.507 P_2 + 55.0 \tag{7.12}$$

7.3 Evaporator System

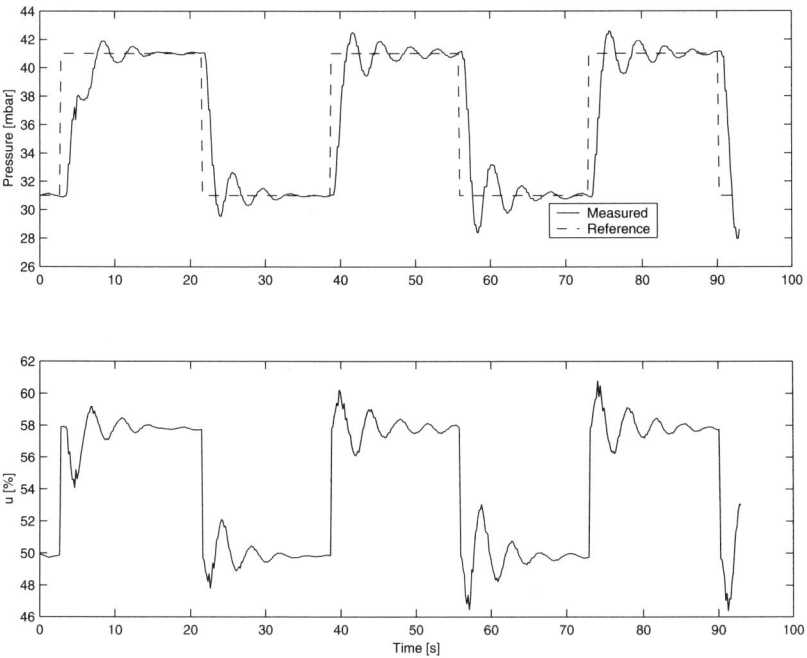

Fig. 7.8. Pressure plant. Trajectories under PI control for a square wave reference with no feedback linearisation

$$F_4 = \frac{1}{\lambda}\left(Q_{100} - F_1 C_P \left(T_2 - T_1\right)\right) \tag{7.13}$$

$$T_{100} = 0.1538 P_{100} + 90.0 \tag{7.14}$$

$$Q_{100} = UA_1\left(T_{100} - T_2\right) \tag{7.15}$$

$$UA_1 = 0.16\left(F_1 + F_3\right) \tag{7.16}$$

$$F_{100} = \frac{Q_{100}}{\lambda_s} \tag{7.17}$$

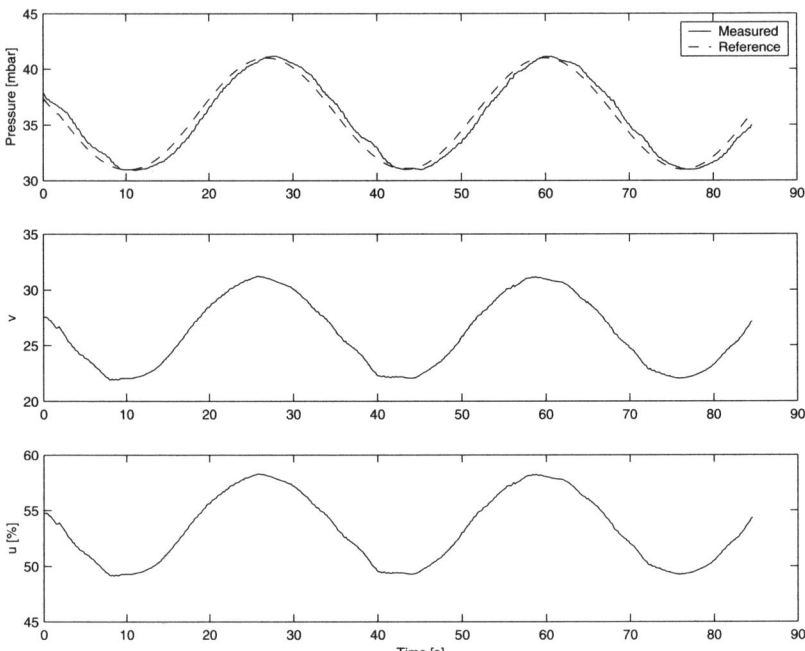

Fig. 7.9. Pressure plant. Globally linearising control trajectories for a sinusoidal reference

$$Q_{200} = UA_2 \left(T_3 - T_{200}\right) \left(1 + \frac{UA_2}{2C_P F_{200}}\right)^{-1} \tag{7.18}$$

$$F_5 = \frac{Q_{200}}{\lambda} \tag{7.19}$$

The model is non-affine with three states. The evaporator variables, their descriptions and normal operating point with measurements units are shown in Table 7.1. The description and value of the evaporator constants are given in Table 7.2

The manipulated variables or inputs are the product flowrate (F_2), the steam temperature (T_{100}) and the coolant flowrate (F_{200}). The variables to be controlled or outputs are the separator level (L_2), the product composition (X_2) and the operating pressure (P_2). Variables F_1, F_3, X_1, T_1 and T_{200} are considered as disturbance sources when they exceed normal operating regions. The dynamics of the single pair separator level (L_2) - product flowrate (F_2) presents an integration in the forward path. F_2 is therefore used in a single gain feedback loop to regulate the level, resulting in a 3-input 3-output bounded-input–bounded-output (BIBO) stable system, as seen in Figure 7.14. The final model for the simulation is given by the same expressions

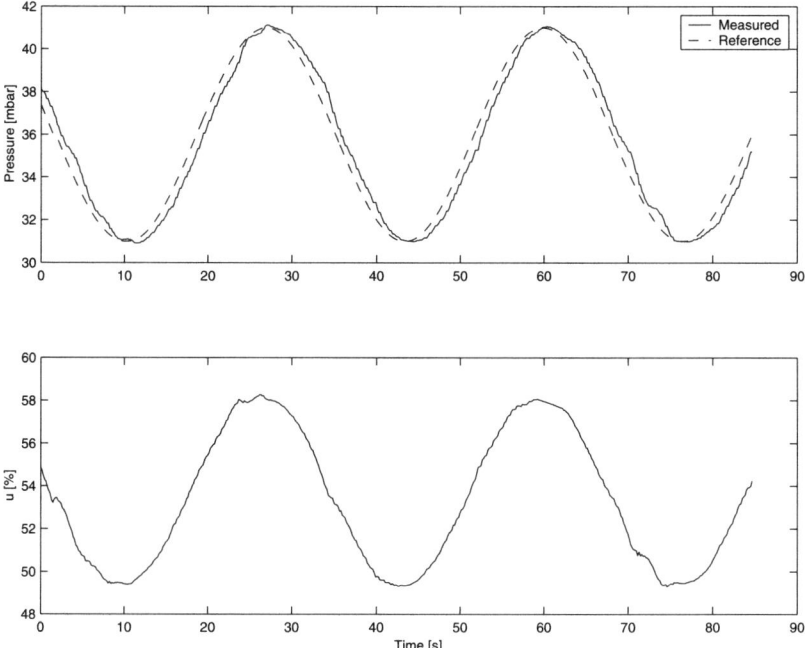

Fig. 7.10. Pressure plant. Trajectories under PI control for a sinusoidal reference with no feedback linearisation

in Equations 7.8–7.19 with the product flowrate F_2 given by

$$F_2 = F_{20} + K_p(L_{2,p} - L_2) \qquad (7.20)$$

where F_{20} is a constant bias, $L_{2,p}$ is the setpoint of L_2 ($L_{2,p}$ is now considered a manipulated variable) and $K_p = -1$ is a proportional gain, which is negative as L_2 decreases when L_2 increases and vice-versa.

In general, the evaporator is represented by a system of the form,

$$\begin{aligned} \dot{x} &= f(x, u) \\ y &= h(x) = x \end{aligned} \qquad (7.21)$$

where x are coordinates on $\Re^3$ and $f(\cdot, \cdot)$ is a smooth vector field.

Process dynamics. As can be seen in Equations 7.8–7.19, the dynamics of the system are highly interconnected. A step test of the nonlinear model was carried out. In the first part of the experiment, positive and negative steps of 0.1 kg/min from the operating point (see Table 7.1) in the product flowrate F_2 were introduced to the system. The responses are shown in Figure 7.15(a). A further test was carried out by introducing a positive and negative step of 1°C in the steam temperature T_{100}. The responses are given in Figure

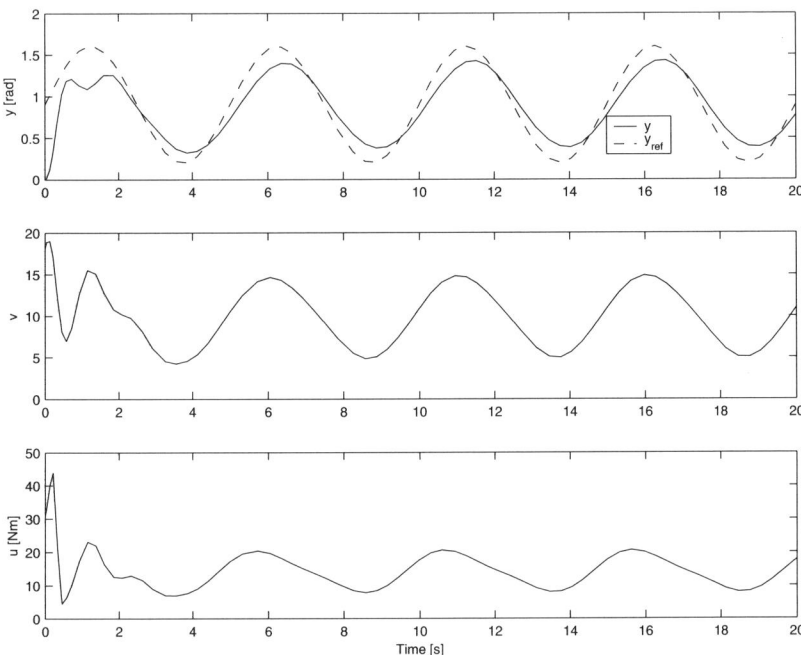

Fig. 7.11. Single link manipulator. Globally linearising control trajectories for an offset sinusoidal reference

7.15(b). Finally, Figure 7.15(c) shows the response of the evaporator states to ±10 kg/min step changes in the water flowrate F_{200}.

The open loop simulations of the step responses show inverted input actions in which positive increments yield to negative responses and vice versa. The process is multivariable, dynamically nonlinear with strong interactions and very sensitive to small input changes. Using a feedback linearisation-decoupling structure would diminish undesired nonlinear connections in the states, facilitating the inclusion of a conventional multivariable linear controller. A linearising feedback control law was computed for a neural network model of the process and applied to the plant. Conventional approaches to control the evaporator system are illustrated in the next section by means of simulation studies.

Single loop control. For any multivariable process, the first control approach to consider is to install one or more single-input single-output control loops, typically using a proportional-integral-derivative (PID) control law. The first step for this approach is to pair the input and output variables to establish the individual control loops. The performance of individual loops is very sensitive to the variables paired. For this purpose, the following section describes the relative gain array method, a conventional approach for

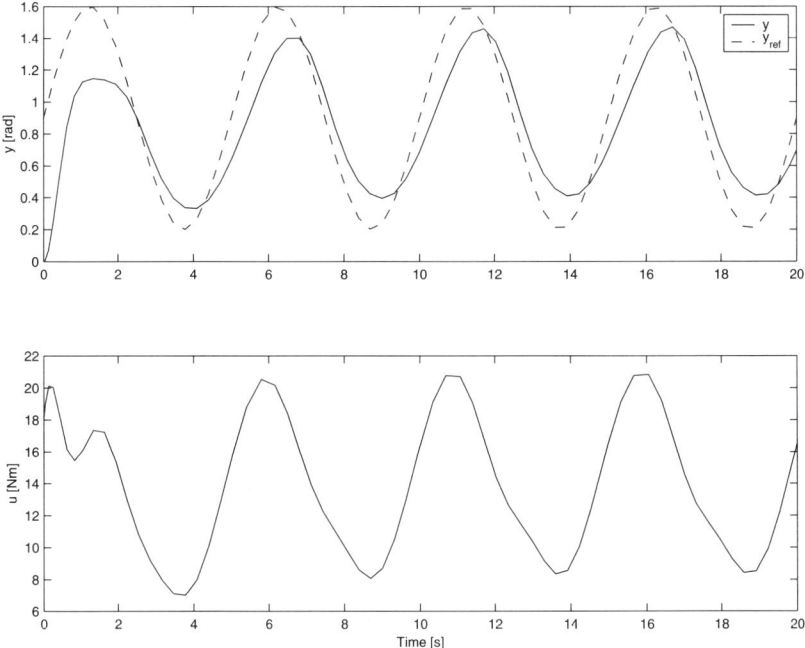

Fig. 7.12. Single link manipulator. Trajectories under PI control only for an offset sinusoidal reference

the selection of input–output pairings based on the static matrix gain of the process.

Relative gain array (RGA). Pairing input and output variables for single loop control is generally not a straightforward task. There are cases in which the pairing is carried out heuristically by considering known cause–effect relationships in the process, such as input flow – level, heat injection – temperature and other self regulatory loops. The RGA technique was first developed in static form using heuristic arguments [133]. Later developments have reviewed alternative methods incorporating dynamic information [134, 1].

The RGA of a nonsingular square gain matrix G is a square matrix defined as

$$G_{RGA} = G .* G^{-T} \qquad (7.22)$$

where '.*' denotes element-wise multiplication of the matrices, and $\{\cdot\}^{-T}$ denotes the transpose of the inverse. The input matrix G holds the gains K_{ij} of the corresponding steady-state process model

$$C_1 = K_{11}M_1 + \ldots + K_{1n}M_n$$
$$\vdots \qquad (7.23)$$
$$C_n = K_{n1}M_1 + \ldots + K_{nn}M_n$$

148   7. Case Studies

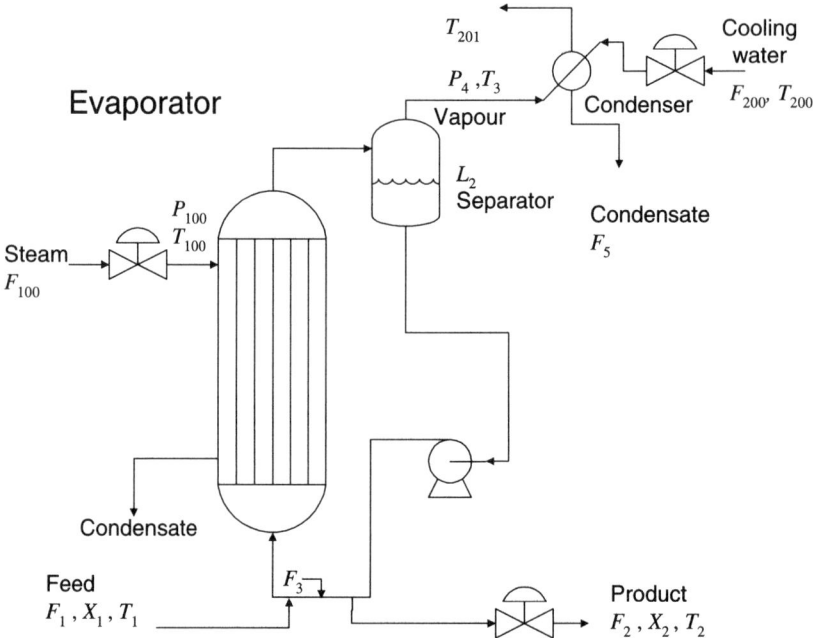

Fig. 7.13. Forced circulation evaporator - process diagram

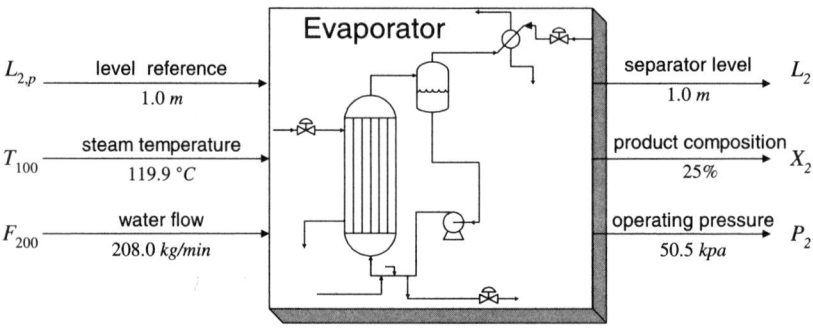

Fig. 7.14. Evaporator diagram

where M_j and C_i are the steady-state values of the ith input and jth output respectively. The element ij in the RGA array shown in Equation 7.22 can be interpreted as the gain from input u_j to output y_i when the other inputs are zero (all other loops open), divided by the corresponding gain when all other outputs are zero (all other loops have maximally tight control).

The recommended one-to-one pairing of input and output variables is given by the larger indexes in the matrix. An RGA element equal to unity indicates the desired pairing of the input and output variable with no inter-

7.3 Evaporator System   149

Table 7.1. Description and operating points of the evaporator variables

Variable	Description	Units	Value
F_1	Feed flowrate	kg/min	10.0
F_2	Product flowrate	kg/min	2.0
F_3	Circulating flowrate	kg/min	50.0
F_4	Vapour flowrate	kg/min	8.0
F_5	Condensate flowrate	kg/min	8.0
X_1	Feed composition	%	5.0
X_2	Product composition	%	25.0
T_1	Feed temperature	°C	40.0
T_2	Product temperature	°C	84.6
T_3	Vapour temperature	°C	80.6
L_2	Separator level	m	1.0
$L_{2,p}$	Internal level reference	m	1.0
P_2	Operating pressure	kPa	50.5
F_{100}	Steam flowrate	kg/min	9.3
T_{100}	Steam temperature	°C	119.9
P_{100}	Steam pressure	kPa	194.7
Q_{100}	Heater duty	kW	339.0
F_{200}	Cooling water flowrate	kg/min	208.0
T_{200}	Cooling water inlet temperature	°C	25.0
T_{201}	Cooling water outlet temperature	°C	46.1
Q_{200}	Condenser duty	kW	307.9

Table 7.2. Description and value of the evaporator constants

Constant	Description	Value
ρ	Liquid density	$\rho A = 20$ kg/m
A	Cross-sectional area of the separator	$\rho A = 20$ kg/m
M	Amount of liquid in the evaporator	20 kg
C	Mass–pressure conversion constant	4 kg/kPa
C_P	Heat capacity of the liquor	0.07 kW/K(kg/min)
λ	Latent heat of vaporisation of the liquor	38.5 kW/(kg/min)
λ_S	Latent heat of steam	36.6 kW/(kg/min)
UA_2	Overall heat transfer coefficient times the heat transfer area	6.84 kW/K

actions. The more an element deviates in value from unity, the stronger the interactions are.

For the evaporator, the loop gains are given by the matrix G, calculated from the open loop test in Figure 7.15.

$$G = \begin{bmatrix} -1 & -0.08 & -0.003 \\ 0.05 & 0.8 & 0.04 \\ 0.1 & 0.8 & 0.06 \end{bmatrix} \tag{7.24}$$

The RGA is evaluated using Equation 7.22 resulting in

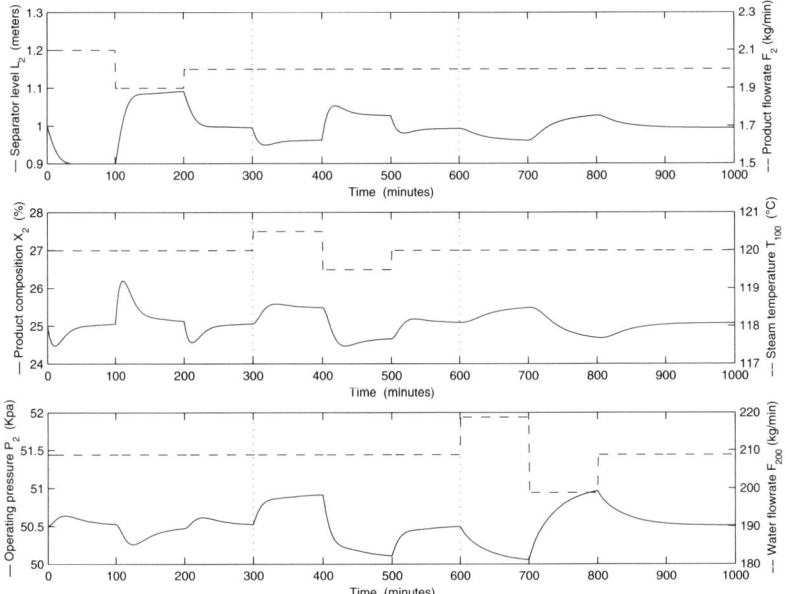

Fig. 7.15. Open loop responses of the states of the evaporator to positive and negative step increments in the inputs. Phase (a) shows the system response to an increment in F_2. In phase (b) the step is introduced in input X_2 and in phase (c) the system responds to step changes in F_{200}. Solid lines: state outputs. Dashed lines: inputs. Phase (a): (0–300 min), phase (b): (300–600 min) and phase (c): (600–1000 min)

$$G_{RGA} = \begin{array}{c} L_2 \\ X_2 \\ P_2 \end{array} \begin{array}{ccc} F_2 & T_{100} & F_{200} \\ \left[\begin{array}{ccc} 1.0028 & 0.0056 & -0.0084 \\ -0.0084 & 2.9916 & -1.9833 \\ 0.0056 & -1.9972 & 2.9916 \end{array}\right] \end{array} \qquad (7.25)$$

From the previous Equation 7.25, the recommended loop pairings are: (i) Product flowrate (F_2) → separator level (L_2), (ii) Steam temperature (T_{100}) → product composition (X_2) and (iii) Coolant flowrate (F_{200}) → operating pressure (P_2).

The matrix in Equation 7.25 indicates strong interconnection in the steady-state response of the plant. It is also viable to pair the loops $X_2 \rightarrow F_{200}$ and $P_2 \rightarrow T_{100}$ but weaker interconnections are generally easier to decouple and the chances of forcing intrinsic links that may harm the closed loop dynamics are smaller. The next section presents a simulation test of conventional PID controllers in the individual loops resulting from the RGA pairing technique.

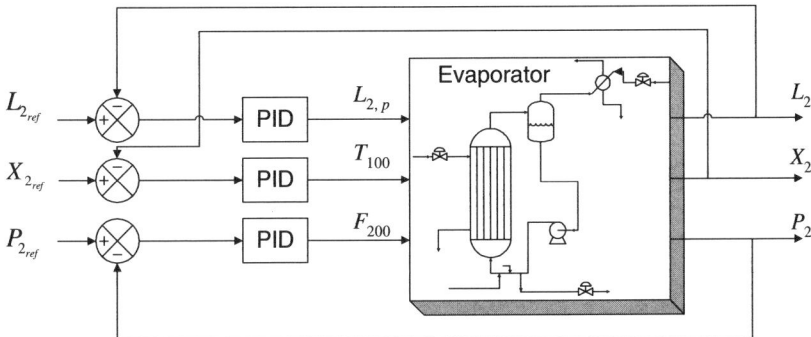

Fig. 7.16. Single loop control diagram for the evaporator system

7.3.1 Single loop control simulations on the evaporator

Three step tests are carried out on the evaporator model using single PID control loops as shown in Figure 7.16. The tuning of the linear controllers is achieved by heuristically changing the gains until satisfactory responses are obtained at the three outputs. Standard PID tuning procedures can alternatively be used to obtain a first estimate. However, the highly nonlinear dynamics and strong connections of the loops demand trial-and-error modification of the gains until the best possible tuning is achieved. Furthermore, the controller performs satisfactorily only inside a very restricted region in the neighbourhood of the operating point for which the controller was tuned.

The results for a 20% positive and negative level step test are shown in Figure 7.17. The response shows a damping ratio of approximately 0.2 and a settling time of more than 100 minutes for the level. With strong interactions and large settling times, the composition and the pressure dynamics do not meet design specifications.

The next test introduces a 12% positive and negative step increment on the product composition (X_2), see Figure 7.18. The composition loop presents a slow response time barely settling in 150 minutes, however, the composition change immediately affects the other two loops reaching overshoots of almost 10% of the operating value.

The results from the final step response test are shown in Figure 7.19. This time increments of only $\pm 3\%$ in the operating pressure setpoint triggers oscillatory responses of more than 10% overshoots. The pressure settles in approximately 80 minutes with an acceptable damping ratio of almost 0.6.

Discussion. PID controllers are very reliable and versatile providing control solutions relatively easily, including design and operability. Nevertheless, as is shown in the evaporator simulations, there are several disadvantages when they are used on multivariable nonlinear processes:

- *Restricted operating regions.* In view of the nonlinearities of the process, different initial conditions lead to significantly different trajectories in the

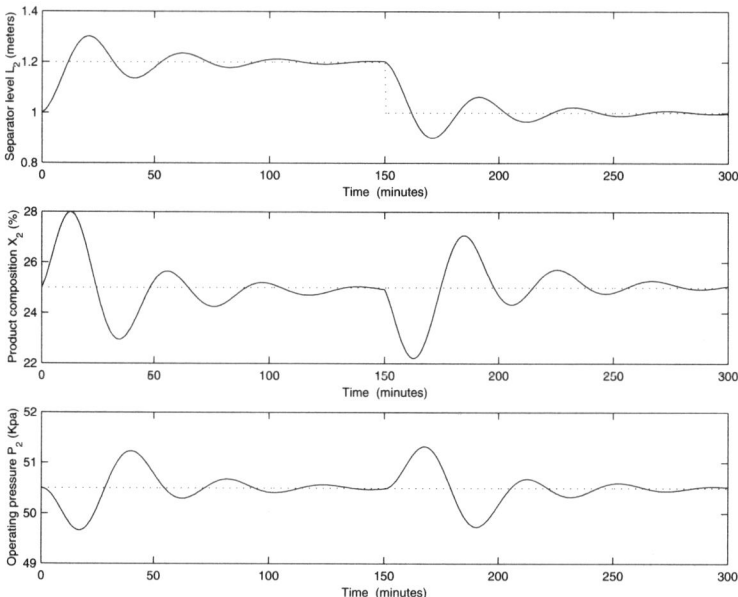

Fig. 7.17. Single loop control of the evaporator. Step response to a 20% positive and negative on the separator level (L_2). Solid line: output variables. Dotted line: setpoints

states and outputs and consequently to small operating ranges for fixed controller gains.
- *Slow responses with large settling times.* To dynamically compensate for undesired dynamic behaviour of the output variables in a reduced time period generally leads to excessively large control signals. Keeping these control variables within the operating ranges of the actuators is commonly detrimental to the response speed of the output dynamics.
- *Retuning is often necessary in the presence of minor changes in operating conditions.* Small changes in the parameters of the plant or the operating points of the variables considered as disturbances, cause the plant dynamics to change.
- *Strongly coupled dynamics.* Standard control techniques do not provide dynamic compensation to process interactions. Strongly coupled outputs are always an unwanted factor in control rooms.

The evaporator system described in this section is used as a case study for the linearising-decoupling feedback designed using a dynamic neural model of the plant. Its multivariable nature and strong nonlinear interconnections

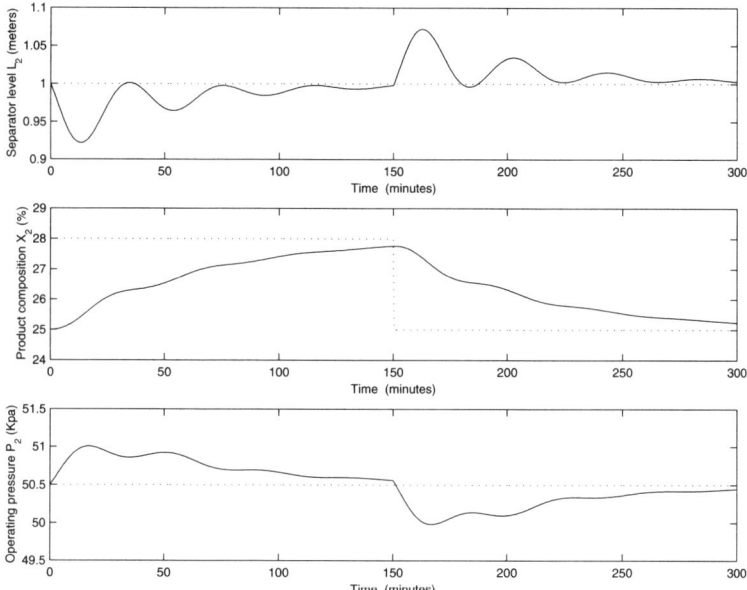

Fig. 7.18. Single loop control of the evaporator. Step response to a 12% positive and negative increases on the product composition (X_2) reference signal. Solid line: output variables. Dotted line: setpoints

make the evaporator system a suitable test bed for the control strategies presented in this book.

7.3.2 Approximate feedback linearisation-decoupling of the evaporator

The feedback linearisation control methodology described in previous chapters was implemented on the evaporator model. The manipulated variables are: the internal reference for the separator level ($L_{2,r}$), steam temperature (T_{100}) and cooling water flowrate (F_{200}), while the outputs of the system are: separator level (L_2), product composition (X_2) and operating pressure (P_2).

Modelling of the evaporator system using dynamic neural networks. The training of the DNN used for the identification of the evaporator system was carried out by means of the combined method proposed in Section 4.6.

The training data was formed by applying steps to the three inputs. The training batch selected includes a step response for the product flowrate F_2 (0–100s), a step response for the steam temperature T_{100} (100–200s) and

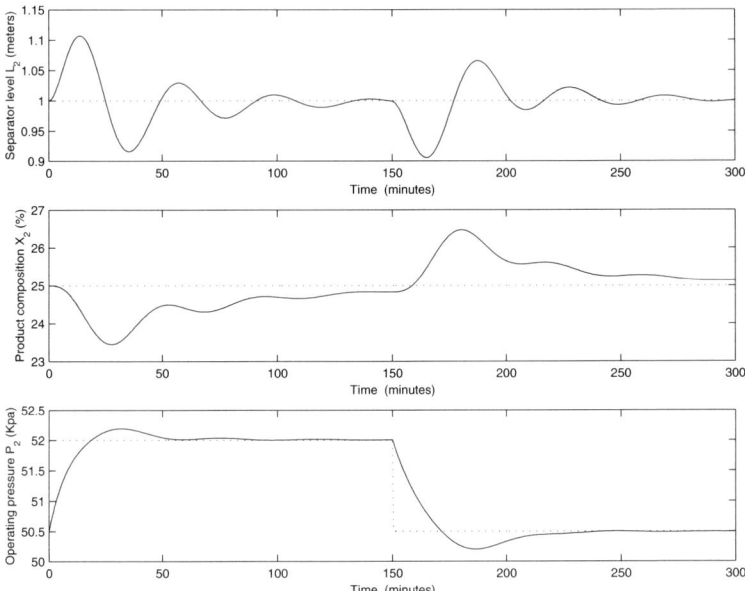

Fig. 7.19. Single loop control of the evaporator. Response to a 3% positive and negative step on the operating pressure (P_2). Solid line: output variables. Dotted line: setpoints

finally the response for step in the cooling water flowrate F_{200}. A periodic square wave was superimposed on the input signals to enhance the excitation. Its amplitude was 10% of the step size and its period much smaller than the minimum time constant of the system. The signals had to be scaled due to the discrepancies in the operating points of the variables. For example, the nominal operating point of the cooling water flowrate is 208 kg/min while the separator level is 1 m. Consequently, the patterns were scaled to fit the interval [0, 1] under the conversion given in Equation 7.26:

$$z_{\text{scaled}} = \frac{z - z_{\min}}{z_{\max} - z_{\min}} \tag{7.26}$$

The network outputs are the first three states $[x_{n_1}, x_{n_2}, x_{n_3}]$ of the output state vector, the number of states in the selected network was greater than the order of the physical model.

Training results and validation. The output vector of the DNN and the training data are plotted in Figure 7.21. The DNN accurately modelled the training set with a significantly small error.

Testing the network in stand-alone operation against a validation input sequence produces the results shown in Figure 7.22. The network not only

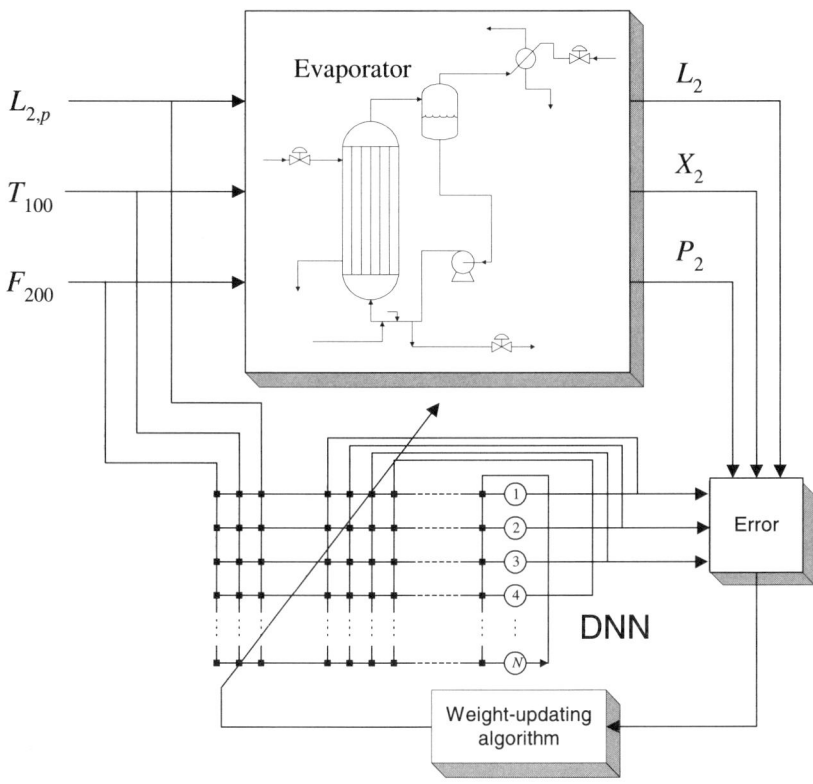

Fig. 7.20. Dynamic neural network for identification of the evaporator system

reproduces the output to step changes in the inputs but more importantly, to random input sequences in all three inputs. Although the vector relative degree of the identified network was {1 1 1}, corresponding to the outputs L_2, X_2 and P_2, respectively, the best results were obtained assuming a vector relative degree {2 2 1} to compute the linearisation and decoupling. See Remark 3.1.4.

Control algorithm for the evaporator system. After the process was identified off-line using a dynamic neural network, the feedback linearisation-decoupling law was computed for the resulting neural model. Figure 7.23 shows the globally linearising control scheme applied to the evaporator. The design of the linearising feedback was accomplished by selecting the poles of the three linearised loops, i.e. selecting the values of the design parameters $\hat{\lambda}_{ik}$s. The characteristic polynomial of the first two loops has two poles while the third has only one. The poles do not speed up the response dramatically but ensure stability, input–output linearity and weak interactions. Having a system with a linear input–output behaviour, a set of PI controllers was introduced to achieve final desired closed-loop dynamics: fast input track-

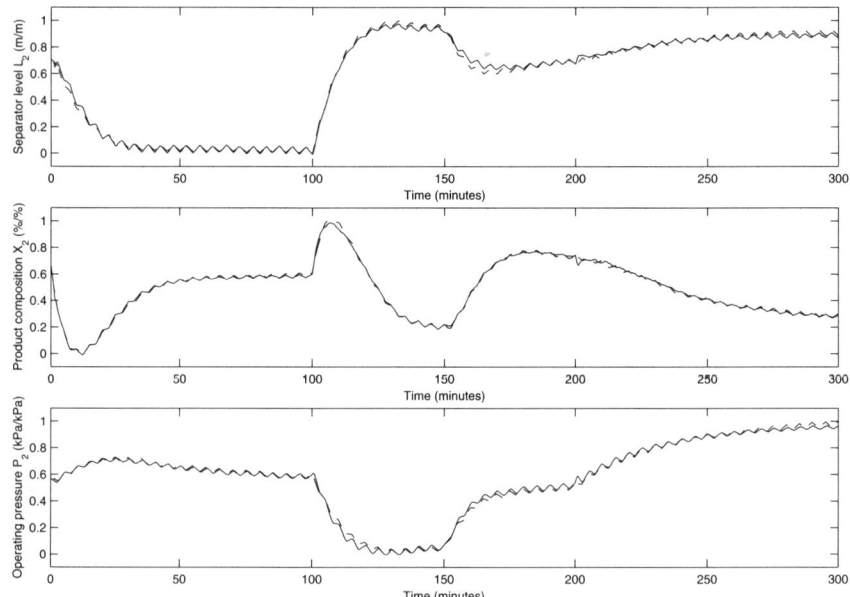

Fig. 7.21. Nonlinear multivariable identification of the evaporator using a dynamic neural network. Solid line: network model; dashed line: training data

ing, stability to moderate step increments on the setpoints and physically realisable control signals.

The poor control resulting from strong process interactions between the pressure control loop and the composition control loop on the evaporator is a well-known difficulty in practical implementation of conventional controllers. The interactions between loops attempting to control both top and bottom compositions on distillation columns is another infamous case of this common problem. The decoupling action of the controller is not intended to completely eliminate the interaction between the output variables. The feedback is calculated based on an approximation of the evaporator system and hence full decoupling is not expected. Nevertheless, as shown in the next section, the interactions are reduced significantly to facilitate incorporation of the linear single looped controllers. The tuning of the PI controllers was done by trial and error, until satisfactory responses were achieved.

Simulation results. All simulations were implemented on SIMULINK® using a fourth order variable step Runge–Kutta method for solving the differential equations [135]. The analysis includes positive and negative step tests as well as disturbance and random responses of the linearising-decoupling feedback including the outer PI control loop.

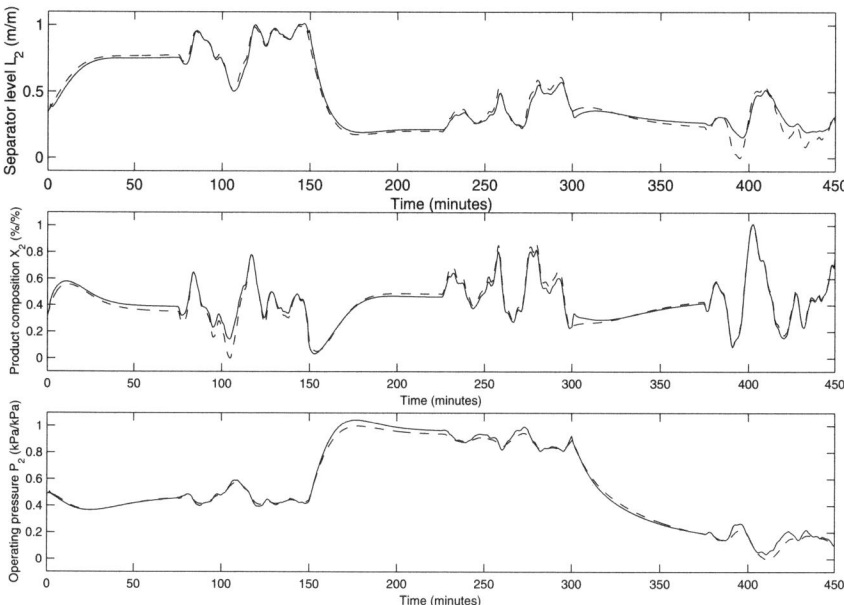

Fig. 7.22. Validation test of the dynamic neural network model of the evaporator. Solid line: network model; dashed line: validation data from the evaporator.

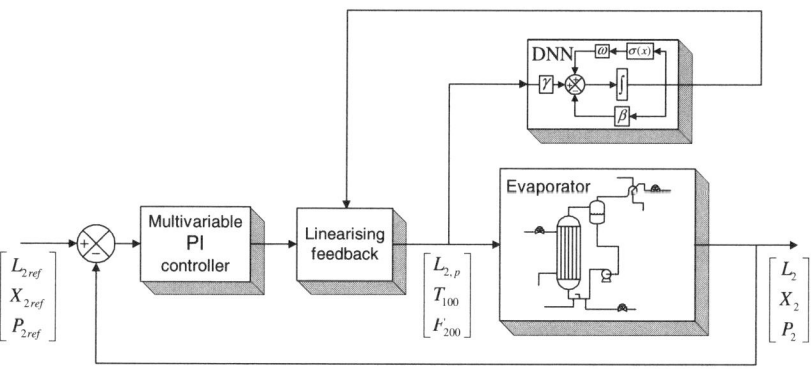

Fig. 7.23. Globally linearising control of the evaporator using a dynamic neural network model

The first test carried out consisted of an increment on the separator level setpoint, see Figure 7.24. In less than 15 minutes, the level settles to a new 20% greater value, at the new setpoint, while the product composition and the pressure do not exceed 2% of their operating points. The solid lines show the output trajectories and the dashed lines the control signals.

158   7. Case Studies

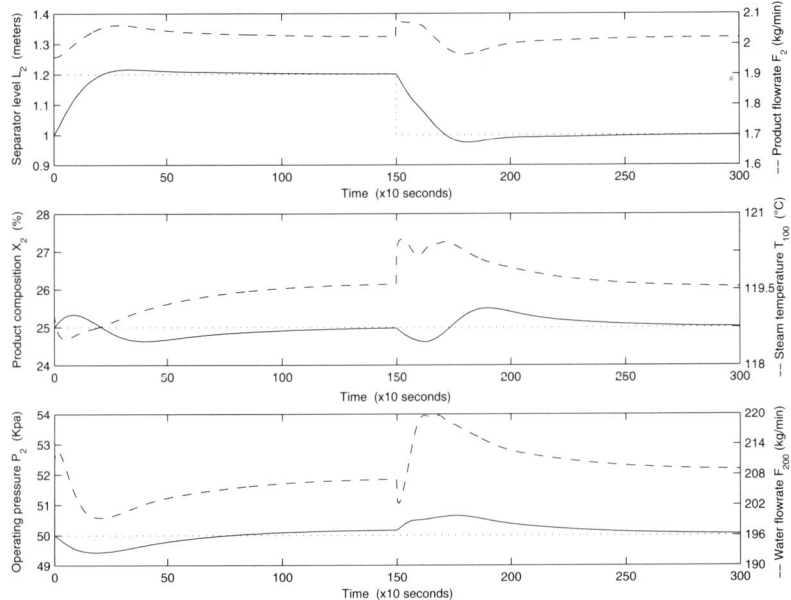

Fig. 7.24. Feedback linearisation and decoupling of the evaporator. Response to ±20% step changes on the separator level (L_2) setpoint. Solid line: output variables. Dashed lines: control variables. Pointed lines: set-point

The control system responds satisfactorily to large step amplitudes in the separator level L_2, keeping the relative interaction margins of the other two loops within a reasonable range. The conventional control approach presented in Figure 7.16 shows much stronger interactions and is not tolerant to large step increments. A second test shows the response of the system to a 10% increase of the product composition X_2 setpoint. Figure 7.25 shows how L_2 and F_{200} keep within 5% and 2%, respectively, of their operating points. Additional simulations not presented here showed that decreasing the PI gains would smooth the control signals, at the expense of increasing the relative interaction margins on the other variables [79]. The decoupling is again not perfect but significantly improves results from conventional approaches as seen in Figure 7.18. For practical purposes, to completely eliminate connections across the outputs is not the objective, as this may require control actions too strong to entirely break intrinsic interactions in the plant. It follows that the interaction margins for the second loop step increments also meets practical design objectives.

The final test shows the responses to ±5% step changes on the operating pressure setpoint, as seen in Figure 7.26. The product composition and

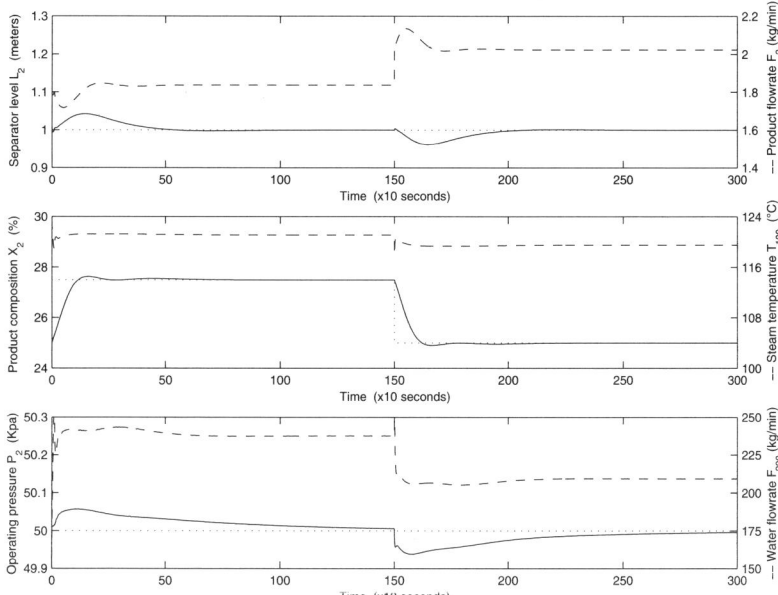

Fig. 7.25. Feedback linearisation and decoupling of the evaporator. Response to ±10% step changes on the product composition (X_2) setpoint. Solid line: output variables. Dashed lines: control variables. Pointed lines: setpoint

the separator level stay within 5% of their operating points. Changes in the manipulated variables are less abrupt than in the previous experiment. The relative gain of the maximum point represents a significant improvement from the open loop performance in which a 1% increment on the operating pressure would lead to a 10% increment of the other variables, as shown in Figure 7.15.

The practical implementation of the control scheme proposed here (see Figure 7.23) would need a 6 × 6 matrix inversion, a hyperbolic tangent calculation and standard algebraic operations repeated for every control sampling step. Only six neurons were necessary to achieve a satisfactory model that was used to compute the linearising laws. The computational cost is therefore within the means of current computer control schemes. The simulations show a satisfactory performance of the linearising control scheme. The speed of response and the strength of the interactions are improved remarkably.

An earlier approach to satisfactorily control the same evaporator model was carried out using a multilayer perceptron for the design of a predictive control law to control [136]. The network is constantly retrained to compensate the rigidity of the model. A practical real time implementation of a

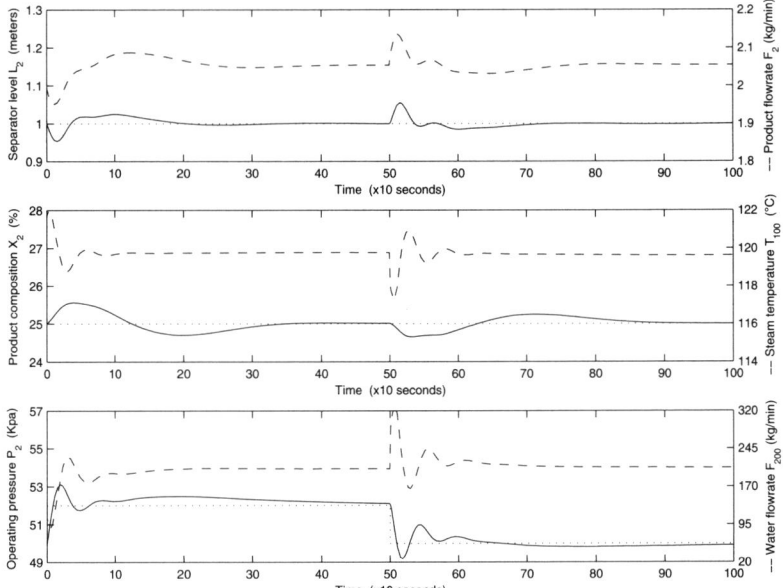

Fig. 7.26. Feedback linearisation and decoupling of the evaporator. Response to a 5% positive and negative step increment on the operating pressure (P_2) setpoint. Solid line: output variables. Dashed lines: control variables. Pointed lines: setpoint

similar MIMO globally linearising feedback method on a real industrial evaporator system was presented in later work [137], where a physical model is used to compute the transformations required for feedback linearisation.

7.4 Summary

In practice, most physical systems are inherently nonlinear and multivariable. Many of these can be controlled adequately by applying linear control theory and conventional PI or PID controllers. Yet, environmental demands, requirements for energy efficiency, product quality, production flexibility, among others, have made process operations more complex, and demand greater operating ranges. In those cases control based on linear methods often do not yield satisfactory results. This chapter presents three case studies: the identification and on-line approximate feedback linearisation of a real single-input single-output pressure pilot plant, the identification and approximate feedback linearisation of a simulated single link manipulator, and the identification, feedback linearisation and decoupling of a 3 × 3 evaporator system.

REFERENCES

1. D. E. Seborg, T. F. Edgar, and D. A. Mellichamp. *Process dynamics and control.* Wiley series in chemical engineering. Wiley, New York, 1989.
2. W. J. Rugh. Analytical framework for gain scheduling. *IEEE Control Systems,* 11:79–84, 1991.
3. J. S. Shamma and M. Athans. Gain scheduling: potential hazards and possible remedies. *IEEE Control Systems,* 12:101–107, 1992.
4. K. J. Åstrom and B. Wittenmark. *Adaptive control.* Addison Wesley, Reading, Massachusetts, 1995.
5. A. Isidori. *Nonlinear control systems.* Springer, Berlin ; New York, 3rd edition, 1995.
6. A. Isidori. *Nonlinear control systems: an introduction.* Springer-Verlag, Berlin; New York, 2nd edition, 1989.
7. C. Kravaris and J. C. Kantor. Geometric methods for nonlinear process-control .1. background. *Industrial & Engineering Chemistry Research,* 29(12):2295–2310, 1990.
8. H. Nijmeijer and A. J. Van der Schaft. *Nonlinear dynamical control systems.* Springer-Verlag, New York, 1990.
9. J. J. E. Slotine and Weiping Li. *Applied nonlinear control.* Prentice Hall, Englewood Cliffs, N.J., 1991.
10. M. A. Henson and D. E. Seborg. *Nonlinear process control.* Prentice Hall, Upper Saddle River, N.J., 1997.
11. C. Kravaris and J. C. Kantor. Geometric methods for nonlinear process-control .2. controller synthesis. *Industrial & Engineering Chemistry Research,* 29(12):2310–2323, 1990.
12. S. Battilotti. *Noninteracting control with stability for nonlinear systems.* Springer-Verlag, London ; New York, 1994.
13. L. Ljung. *System identification: theory for the user.* Prentice-Hall, Englewood Cliffs, NJ, 1987.
14. S. S. Haykin. *Neural networks: a comprehensive foundation.* Prentice Hall, Upper Saddle River, N.J., 2nd edition, 1999.
15. W. S. McCulloch and W. Pits. A logical calculus of the ideas immanent in nervous activity. *Bulletin of Mathematical Biophysics,* 9:127–147, 1943.
16. D. E. Rumelhart and J. L. McClelland. *Parallel distributed processing : explorations in the microstructure of cognition.* MIT Press, Cambridge, Mass., 1986.
17. R. P. Lippmann. An introduction to computing with neural nets. *IEEE ASSP,* pages 4–22, April 1987.
18. P. S. Churchland, T. J. Sejnowski, and T.J. Seynowski. *The computational brain.* MIT Press, Cambridge, Mass., 1992.
19. D. S. Broomhead and D. Lowe. Multi-variable functional interpolation and adaptive networks. *Complex Systems,* 2:321–355, 1988.

20. D. F. Specht. Probabilistic neural networks. *Neural Networks*, 3(1):109–118, 1990.
21. J. A. Leonard and M. A. Kramer. Radial basis functions networks for classifying process faults. *Control Systems*, 11:31–38, 1991.
22. J. A. Leonard, M. A. Kramer, and L. H. Ungar. Using radial basis functions to approximate a function and its error-bounds. *IEEE Transactions on Neural Networks*, 3(4):624–626, 1992.
23. J. Albus. A new approach to manipulate control: The cerebellar model articulation controller. *Transactions of ASME Journal of Dynamical Systems, Measurement and Control*, 97:220–227, 1975.
24. W. T. Miller, F. H. Glanz, and L. G. Kraft. Cmac - an associative neural network alternative to backpropagation. *Proceedings of the IEEE*, 78(10):1561–1567, 1990.
25. J. J. Hopfield. Neural networks and physical systems with emergent collective computational abilities. *Proceedings of the National Academy of Sciences of the United States of America-Biological Sciences*, 79(8):2554–2558, 1982.
26. P. Koiran. Dynamics of discrete-time, continuous state hopfield networks. *Neural Computation*, 6(3):459–468, 1994.
27. D. Ackley, G. E. Hinton, and T. J. Sejnowski. A learning algorithm for Boltzmann machines. *Cognitive Science*, 9:147–169, 1982.
28. E. H. L. Aarts and J. H. M. Korst. Boltzmann machines and their applications. *Lecture Notes in Computer Science*, 258:34–50, 1987.
29. T. Kohonen. *Self-organization and associative memory*. Springer-Verlag, Berlin; New York, 3rd edition, 1989.
30. S. Grossberg. Adaptive pattern classification and universal recording: I. parallel development and coding of neural feature detectors. *Biological Cybernetics*, 23:123–129, 1976.
31. L. Ljung. *System identification: theory for the user*. Prentice Hall PTR, Upper Saddle River, NJ, 2nd edition, 1999.
32. O. Nelles. *Nonlinear system identification*. Springer-Verlag, Berlin, 2001.
33. K. S. Narendra and K. Parthasarathy. Identification and control of dynamical systems using neural networks. *IEEE Transactions on Neural Networks*, NN1:4–27, 1990.
34. M. Agarwal. A systematic classification of neural-network-based control. *IEEE Control Systems Magazine*, 17(2):75–93, 1997.
35. M. Norgaard, O. Ravn, N. K. Poulsen, and L. K. Hansen. *Neural networks for modelling and control of dynamic systems*. Springer, London, 2000.
36. K. J. Hunt and D. Sbarbaro-Hofer. Neural networks for nonlinear internal model control. *IEE Proceedings – Control Theory and Applications*, 138:431–438, 1991.
37. A. Yesildirek and F. L. Lewis. Feedback linearization using neural networks. *Automatica*, 31(11):1659–1664, 1995.
38. S. L. He, K. Reif, and R. Unbehauen. A neural approach for control of nonlinear systems with feedback linearization. *IEEE Transactions on Neural Networks*, 9(6):1409–1421, 1998.
39. J. A. K. Suykens, B. L. R. Demoor, and J. Vandewalle. Nonlinear-system identification using neural state-space models, applicable to robust-control design. *International Journal of Control*, 62(1):129–152, 1995.
40. A. M. Shaw and F. J. Doyle. Multivariable nonlinear control applications for a high purity distillation column using a recurrent dynamic neuron model. *Journal of Process Control*, 7(4):255–268, 1997.

41. C. Kambhampati, R. J. Craddock, M. Tham, and K. Warwick. Inverse model control using recurrent networks. *Mathematics and Computers in Simulation*, 51(3-4):181–199, 2000.
42. A. Delgado, C. Kambhampati, and K. Warwick. Input/output linearization using dynamic recurrent-neural networks. *Mathematics and Computers in Simulation*, 41(5-6):451–460, 1996.
43. S. Jagannathan, S. Commuri, and F. L. Lewis. Feedback linearization using cmac neural networks. *Automatica*, 34(5):547–557, 1998.
44. H. A. B. Braake, E. J. L. Van Can, J. M. A. Scherpen, and H. B. Verbruggen. Control of nonlinear chemical processes using neural models and feedback linearization. *Computers and Chemical Engineering*, 22(7/8):1113–1127, 1998.
45. M. A. Brdys and G. J. Kulawski. Dynamic neural controllers for induction motor. *IEEE Transactions on Neural Networks*, 10(2):340–355, 1999.
46. D. M. Prett and R. D. Gillette. Optimisation and constrained multivariable control of a catalytic cracking unit. In *AIChE 86th National Meeting*, Houston, Texas, 1979.
47. J. Qin. An overview of industrial model-based predictive control. In *Chemical Process Control V - AIChE Symposium Series*, volume 93, pages 232–256, 1997.
48. D. J. Sandoz, B. Lennox, P. R. Goulding, P. J. Thorpe, T. Kurth, M. J. Desforges, and I. S. Woolley. Innovation in industrial model predictive control. *Computing and Control Engineering Journal*, 10(5):189–197, 1999.
49. B. W. Bequette. Nonlinear control of chemical processes - a review. *Industrial & Engineering Chemistry Research*, 30(7):1391–1413, 1991.
50. J. W. Eaton and J. B. Rawlings. Feedback-control of chemical processes using online optimization techniques. *Computers & Chemical Engineering*, 14(4-5):469–479, 1990.
51. P. B. Sistu, R. S. Gopinath, and B. W. Bequette. Computational issues in nonlinear predictive control. *Computers & Chemical Engineering*, 17(4):361–366, 1993.
52. S. L. Oliveira, V. Nevistic, and M. Morari. Control of nonlinear systems subject to input constraints. In *Proceedings IFAC Nonlinear Control Systems Design Symposium*, pages 15–20, Tahoe City, California, 1995.
53. V. Nevistic and Primbs. J. A. Model predictive control: breaking through constraints. In *Proceedings of the 35th Conference on Decision and Control*, pages 3932–3937, Kobe, Japan, 1996.
54. M. Soroush and H. M. Soroush. Input–output linearizing nonlinear model predictive control. *International Journal of Control*, 68(6):1449–1473, 1997.
55. M. Ayala-Botto, T. Van Den Boom, A. Krijgsman, and J. Sa da Costa. Predictive control based on neural network models with i/o feedback linearization. *International Journal of Control*, 72(17):1538–1554, 1999.
56. H.K. Khalil. *Nonlinear systems*. Prentice Hall, Upper Saddle River, NJ, 2nd edition, 1996.
57. W. Hahn. *Stability of motion*. Springer-Verlag, Berlin, 1967.
58. S. Sastry and M. Bodson. *Adaptive control: stability, convergence and robustness*. Prentice Hall, New Jersey, 1989.
59. J. C. Kantor. An overview of nonlinear geometrical methods for process control. In D.M. Prett and M. Morari, editors, *Shell process control workshop*. Butterworths, New York, 1987.
60. R. M. Hirschorn. Invertibility of multivariable nonlinear control systems. *IEEE Transactions on Automatic Control*, AC-24:855–857, 1979.
61. I. J. Ha and E. G. Gilbert. A complete characterization of decoupling control laws for a general-class of nonlinear-systems. *IEEE Transactions on Automatic Control*, 31(9):823–830, 1986.

62. C. H. Moog. Nonlinear decoupling and structure at infinity. *Mathematics of Control, Signals and Systems*, 1:257–262, 1988.
63. C. Kravaris and M. Soroush. Synthesis of multivariable nonlinear controllers by input output linearization. *AIChE Journal*, 36(2):249–264, 1990.
64. B. Noble and J.W. Daniel. *Applied linear algebra*. Prentice-Hall, Englewood Cliffs, N.J., 1988.
65. C. E. Garcia and M. Morari. Internal model control .3. multivariable control law computation and tuning guidelines. *Industrial & Engineering Chemistry Process Design and Development*, 24(2):484–494, 1985.
66. C. T. Chen. *Linear systems theory and design*. Oxford University Press, New York, 1999.
67. F. C. Chen and C. C. Liu. Adaptively controlling nonlinear continuous-time systems using multilayer neural networks. *IEEE Transactions on Automatic Control*, 39(6):1306, 1994.
68. A. Isidori and J. W. Grizzle. Fixed modes and nonlinear noninteracting control with stability. *IEEE Transactions on Automatic Control*, 33(10):907 914, 1988.
69. H.K. Khalil. *Nonlinear systems*. Macmillan, New York, 1992.
70. D. Claude, M. Fliess, and A. Isidori. Direct and feedback immersion of a nonlinear system in a linear-system. *Comptes Rendus De L Academie Des Sciences Serie I-Mathematique*, 296(4):237–240, 1983.
71. A. Isidori and A. Ruberti. On the synthesis of linear input output responses for nonlinear-systems. *Systems & Control Letters*, 4(1):17–22, 1984.
72. M. A. Henson and D. E. Seborg. Input-output linearization of general nonlinear processes. *AIChE Journal*, 36(11):1753–1757, 1990.
73. A. J. Van der Schaft. Linearization and input-output decoupling for general nonlinear-systems. *Systems & Control Letters*, 5(1):27–33, 1984.
74. B. de Jager. Symbolic aids for modelling, analysis and synthesis of nonlinear control systems. In N. Munro, editor, *Symbolic methods for control systems analysis and design*. The Institution of Electrical Engineers, Stevenage, 1999.
75. H. G. Kwatny and G. L. Blankenship. *Nonlinear Control and Analytical Mechanics*. Birkhäuser, Boston, Mass., 2000.
76. K. J. Hunt, D. Sbarbaro, R. Zbikowski, and P. J. Gawthrop. Neural networks for control-systems - a survey. *Automatica*, 28(6):1083–1112, 1992.
77. K. Funahashi. On the approximate realization of continuous-mappings by neural networks. *Neural Networks*, 2(3):183–192, 1989.
78. F. Garces, C. Kambhampati, and K. Warwick. Dynamic recurrent neural networks for feedback linearization of a multivariable nonlinear evaporator system. In *International congress on intelligent systems and applications, ISA2000*, pages 135–141, Wollongong – Australia, 2000.
79. F. Garces. *Dynamic neural networks for approximate input–output linearisation–decoupling of dynamic systems*. PhD Thesis, University of Reading, 2000.
80. D. O. Hebb. *The organization of behavior; a neuropsychological theory*. Wiley, New York, 1949.
81. F. Rosemblatt. The perceptron: a probabilistic model for information storage and organization in the brain. *Psychological Review*, 65:386–408, 1958.
82. M.L. Minsky and S. Papert. *Perceptrons: an introduction to computational geometry*. MIT Press, Cambridge, Mass.,, 1969.
83. H. T. Siegelmann. *Neural networks and analog computation: beyond the Turing limit*. Birkhauser, Boston, Mass., 1999.
84. J. J. Hopfield. Neurons with graded response have collective computational properties like those of 2-state neurons. *Proceedings of the National Academy*

of Sciences of the United States of America-Biological Sciences, 81(10):3088–3092, 1984.
85. G. E. Hinton. Learning distributed representations of concepts. In *Eighth Annual Conference of the Cognitive Science Society*, Amherst, Mass, 1986.
86. T. J. Sejnowski, P. K. Kienker, and G. E. Hinton. Learning symmetry groups with hidden units - beyond the perceptron. *Physica D*, 22(1-3):260–275, 1986.
87. C. Kambhampati, S. Manchanda, A. Delgado, G. G. R. Green, K. Warwick, and M. T. Tham. The relative order and inverses of recurrent networks. *Automatica*, 32(1):117–123, 1996.
88. A. N. Michel, J. A. Farrell, and W. Porod. Qualitative-analysis of neural networks. *IEEE Transactions on Circuits and Systems*, 36(2):229–243, 1989.
89. W. S. Massey. *A basic course in algebraic topology*. Springer, New York, 3rd edition, 1997.
90. Z. H. Guan, G. Chen, and Y. Qin. On equilibria, stability and instability of Hopfield neural networks. *IEEE Transactions on Neural Networks*, 11:534–540, 2000.
91. K. Matsuoka. Stability conditions for nonlinear continuous neural networks with asymmetric connection weights. *Neural Networks*, 5(3):495–500, 1992.
92. E. N. Sanchez and J. P. Perez. Input-to-state stability (ISS) analysis for dynamic neural networks. *IEEE Transactions on Circuits and Systems I-Fundamental Theory and Applications*, 46(11):1395–1398, 1999.
93. A. Guez, V. Protopopsecu, and J. Barhen. On the stability, storage capacity, and design of nonlinear continuous neural networks. *IEEE Transactions on Systems Man and Cybernetics*, 18(1):80–87, 1988.
94. M. W. Hirsch. Convergent activation dynamics in continuous-time networks. *Neural Networks*, 2(5):331–349, 1989.
95. D. G. Kelly. Stability in contractive nonlinear neural networks. *IEEE Transactions on Biomedical Engineering*, 37(3):231–242, 1990.
96. J. C. Juang. Stability analysis of Hopfield-type neural networks. *IEEE Transactions on Neural Networks*, 10:1366–1374, 1999.
97. S. Hu and J. Wang. Global asymptotic stability and global exponential stability of continuous-time recurrent neural networks. *IEEE Transactions on Automatic Control*, 47:802–807, 2002.
98. E. D. Sontag. On the input to state stability property. *European Journal of Control*, 1:1–24, 1995.
99. E. D. Sontag and Y. Wang. On characterisations of the input-to-state stability property. *Systems and Control Letters*, 24:351–359, 1995.
100. E. Kaszkurewicz and A. Bhaya. On a class of globally stable neural circuits. *IEEE Transactions on Circuits and Systems I-Fundamental Theory and Applications*, 41(2):171–174, 1994.
101. M. Forti, S. Manetti, and M. Marini. Necessary and sufficient condition for absolute stability of neural networks. *IEEE Transactions on Circuits and Systems I-Fundamental Theory and Applications*, 41(7):491–494, 1994.
102. T. Söderström and P. Stoica. *System identification*. Prentice Hall, New York, 1989.
103. V.M. Becerra, J.M.F. Calado, P. Silva, and F. Garces. System identification using dynamic neural networks: training and initialisation aspects. In *Proceedings of the 2002 IFAC World Congress*, Barcelona, Spain, 2002.
104. R. Fletcher. *Practical methods of optimisation*. Wiley, Chichester, 2nd edition, 1987.
105. P. Gill, Murray W., and M. H. Wright. *Practical optimisation methods*. Academic Press, London, 1981.

106. R. L. Haupt and S. E. Haupt. *Practical genetic algorithms.* Wiley, New York, 1998.
107. C. Kambhampati, F. Garces, and K. Warwick. Approximation of non-autonomous dynamic systems by continuous time recurrent neural networks. In S. I. Amari, editor, *Neural networks*, IEEE International Conference on Neural Networks, pages I–64–I–69, Como, Italy, 2000.
108. S. A. Billings and Q. M. Zhu. Nonlinear model validation using correlation tests. *International Journal of Control*, 60(6):1107–1120, 1994.
109. S. A. Billings. Identification of nonlinear systems – a survey. *IEE Proceedings – Control Theory and Applications*, 127:272–285, 1980.
110. S. Chen and S. A. Billings. Neural networks for nonlinear dynamic system modeling and identification. *International Journal of Control*, 56(2):319–346, 1992.
111. A. M. Zoubir and B. Boashash. Nonlinear-system identification - an overview. *Arabian Journal for Science and Engineering*, 18(4):423–458, 1993.
112. B. R. Haynes and S. A. Billings. Global analysis and model validation in nonlinear-system identification. *Nonlinear Dynamics*, 5(1):93–130, 1994.
113. K. Warwick, G. W. Irwin, and K. J. Hunt. *Neural networks for control and systems.* P. Peregrinus on behalf of the Institution of Electrical Engineers, London, 1992.
114. J. Y. Choi, H. F. VanLandingham, and S. Bingulac. A constructive approach for nonlinear system identification using multilayer perceptrons. *IEEE Transactions on Systems Man and Cybernetics Part B- Cybernetics*, 26(2):307–312, 1996.
115. S. H. Tan, J. B. Hao, and J. Vandewalle. Efficient identification of rbf neural-net models for nonlinear discrete-time multivariable dynamical-systems. *Neurocomputing*, 9(1):11–26, 1995.
116. K. Warwick. Neural networks for control - counter arguments. In *International Conference on Control 94, Vols 1 and 2*, pages 95–99, Coventry, UK, 1994.
117. C. Q. Zhang and M. S. Fadali. Nonlinear system identification using a Gabor/Hopfield network. *IEEE Transactions on Systems Man and Cybernetics Part B- Cybernetics*, 26(1):124–134, 1996.
118. S. Adwankar and R. N. Banavar. A recurrent network for dynamic system identification. *International Journal of Systems Science*, 28(12):1239–1250, 1997.
119. K. Funahashi and Y. Nakamura. Approximation of dynamical-systems by continuous-time recurrent neural networks. *Neural Networks*, 6(6):801–806, 1993.
120. M. Kimura and R. Nakano. Learning dynamical systems by recurrent neural networks from orbits. *Neural Networks*, 11(9):1589–1599, 1998.
121. G. Cybenko. Approximation by superpositions of sigmoidal functions. *Mathematics of Control, Signals and Systems*, 2:303–314, 1989.
122. M. W. Hirsch and S. Smale. *Differential equations, dynamical systems and linear algebra.* Academic Press, San Diego, CA, 1974.
123. T. W. S. Chow and X. D. Li. Modeling of continuous time dynamical systems with input by recurrent neural networks. *IEEE Transactions on Circuits and Systems–I: Fundamentals*, 47:575–578, 2000.
124. A. S. Poznyak, W. Yu, E. N. Sanchez, and J. P. Perez. Stability analysis of dynamic neural control. *Expert Systems with Applications*, 14(1-2):227–236, 1998.
125. A. S. Poznyak and E. N. Sanchez. Nonlinear systems approximation by neural networks: error stability analysis. *Intelligent Automation and Soft Computing*, 1:247–258, 1995.

126. G. A. Rovithakis and M. A. Christodoulou. Direct adaptive regulation of unknown nonlinear dynamical- systems via dynamic neural networks. *IEEE Transactions on Systems Man and Cybernetics*, 25(12):1578–1594, 1995.
127. G. F. Franklin, J. D. Powell, and A. Emami-Naeini. *Feedback control of dynamic systems*. Addison-Wesley, Reading, Mass., 3rd edition, 1994.
128. Y. Yasuda. Feedback realizations in linear-multivariable systems - model-following with stability and disturbance rejection. *International Journal of Control*, 42(5):1049–1070, 1985.
129. F.H. Raven. *Automatic control engineering*. McGraw-Hill, New York, 5th edition, 1995.
130. G. A. Rovithakis and M. A. Christodoulou. Adaptive-control of unknown plants using dynamical neural networks. *IEEE Transactions on Systems Man and Cybernetics*, 24(3):400–412, 1994.
131. P. M. Goodall. *The efficient use of steam*. IPC Science and Technology Press, London, UK, 1980.
132. R. B. Newell and P. L. Lee. *Applied process control : a case study*. Prentice-Hall, Englewood Cliffs, N.J., 1989.
133. E. H. Bristol. On a new measure of interaction for multivariable process control. *IEEE Transactions on Automatic Control*, AC-11:133–134, 1966.
134. R. D. Johnston and G. W. Barton. Structural equivalence and model-reduction. *International Journal of Control*, 41(6):1477–1491, 1985.
135. J. R. Dormand and P. J. Prince. A family of embedded Runge-Kutta formulae. *Journal of Computational and Applied Mathematics*, 6:19–26, 1980.
136. P. M. Mills, A. Y. Zomaya, and M. O. Tade. Adaptive model-based control using neural networks. *International Journal of Control*, 60(6):1163–1192, 1994.
137. L. C. To, M. O. Tade, and G. P. Le Page. Implementation of a differential geometric nonlinear controller on an industrial evaporator system. *Control Engineering Practice*, 6(11):1309–1319, 1998.

INDEX

Approximate input–output linearisation, 58
- control affine systems, 122
- general nonlinear systems, 130
Approximation of autonomous nonlinear systems, 104
Approximation of non-autonomous nonlinear systems, 109
Artificial neural networks (ANNs), 61

BFGS training algorithm, 86

Characteristic matrix, 30
Combined training algorithms, 90
Constrained optimisation, 86
Control affine systems, 27
Converse Lyapunov theorem, 26
Covector field, 17
Cross-validation, 93

Decrescent functions, 24
Direct inverse control, 10
Distribution
- complete integrability, 21
- involutive, 20
Distributions, 20
Disturbance rejection, 127
Dynamic neural networks, 61, 66
- approximation ability, 103
- for nonlinear identification, 10
- for nonlinear system approximation, 101
- generalisation capabilities, 61
- initialisation, 82
- initialisation procedure, 82
- input-to-state stability, 73
- overfitting, 91
- overtraining, 91
- parallel-distributed processing, 61
- stability, 69, 72
- structure, 102
- training, 78, 81
- transfer function, 63
- validation, 91
- vector relative degree, 68
Dynamic recurrent neural networks, 11

Equilibrium point, 22
Euclidean space, 15
Evaporator system, 141
- approximate feedback linearisation-decoupling, 153
- control algorithm, 155
- mathematical model, 142
- modelling, 153
- process dynamics, 145
- simulations, 156
- single loop control , 146
- single loop simulations, 151
Exact linearisation, 54
External control, 125

Feedback linearisation, 2, 4
- approximate input–output, 58
- introduction, 27
- using dynamic neural networks, 121
Feedback linearisation-decoupling, 4
Fixed point, 70
Frobenius theorem, 21, 34

Gain scheduling, 2
General nonlinear systems, 57, 130
Genetic Algorithm, 89
Genetic algorithms, 87
- chromosome, 87
- crossover, 87
- for training neural networks, 89
- mutation, 89
- operators, 87
- replication, 87
Geometric theory, 15
- elementary concepts, 15
Gradient based optimisation methods, 82

Index

Gradient descent method, 83

Input–output linearisation, 39
- and decoupling, 46
Input–output linearisation-decoupling, 12
Internal model control, 10

Jacobian linearisation, 2

Lie
- bracket, 19
- derivative, 18, 57
- differential operations, 18
- product, 19
Line search methods, 84
Linear vector spaces, 15
Linearisation by immersion, 55
Lipschitz condition, 21

Matrix norms, 16
Model predictive control (MPC), 12
Model structure selection, 93
Model validation, 8
Multi-input multi-output system (MIMO), 29, 39
Multilayer perceptron, 5

Neural networks, 5
- activation function, 6, 63, 65
- backpropagation, 7, 64
- CMAC networks, 12
- multilayer feedforward networks, 65
- multilayer perceptron, 5, 65
- multilayer perceptrons, 9
- origins, 62
- static neural networks, 9
- structure, 62
- transfer function, 63
- universal approximation property, 101
Newton's method, 85
Nonlinear control
- conventional strategies, 1
- justification, 1
Nonlinear predictive control, 12
Nonlinear system models, 4
Nonlinear systems
- control affine systems, 27
- general, 57
- stability, 21
normal form, 33

PI control, 126

Positive definite functions, 24
Pressure pilot plant, 135
- description, 135
- globally linearising control, 137
- modelling, 136

Quasi-Newton method, 85

Random search training algorithms, 87
Regularisation, 93
Relative degree, 30, 57
- total, 30
- vector, 30, 32
Relative gain array (RGA), 147

Sequential quadratic programming, 86
Sets, 17
- boundary, 17
- bounded sets, 17
- closed sets, 17
- compact set, 17
- convex set, 17
- interior and closure, 17
- open sets, 17
Single link manipulator
- approximate feedback linearisation, 138
- dynamic neural network model, 94
Stability, 21
- asymptotic, 23
- exponential, 23
- in the sense of Lyapunov, 22, 24, 25
- input-to-state, 74
- internal, 49
- of input–output linearised systems, 48
- region of attraction, 23
- zero dynamics, 39
Stability analysis of feedback linearised systems, 128
Stability under external PI control, 130
Steepest descent method, 84
Subspace, 15
System identification, 8
- model structure selection, 8
- model validation, 8
- parameter estimation, 8

Universal approximation property of neural networks, 101

Vector field, 17
- covector field, 17
- differential operations, 18

Vector norms, 16
Volterra linearisation, 56

Zero dynamics, 33
– stability, 39